Springer Series in Materials Science

Volume 249

The Springer Series in Materials Science covers the complete spectrum of materials physics, including fundamental principles, physical properties, materials theory and design. Recognizing the increasing importance of materials science in future device technologies, the book titles in this series reflect the state-of-the-art in understanding and controlling the structure and properties of all important classes of materials.

More information about this series at http://www.springer.com/series/856

Ji-Guang Zhang · Wu Xu
Wesley A. Henderson

Lithium Metal Anodes and Rechargeable Lithium Metal Batteries

 Springer

Ji-Guang Zhang
Energy and Environment Directorate
Pacific Northwest National Laboratory
Richland, WA
USA

Wesley A. Henderson
U.S. Army Research Office (ARO)
Research Triangle Park, NC
USA

Wu Xu
Energy and Environment Directorate
Pacific Northwest National Laboratory
Richland, WA
USA

ISSN 0933-033X ISSN 2196-2812 (electronic)
Springer Series in Materials Science
ISBN 978-3-319-82971-5 ISBN 978-3-319-44054-5 (eBook)
DOI 10.1007/978-3-319-44054-5

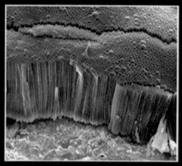

Lithium Metal
Anodes and
Rechargeable
Lithium
Metal Batteries

Ji-Guang Zhang
Wu Xu
Wesley A. Henderson

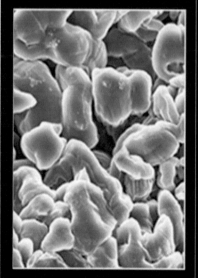

Li Metal & Anode | Batteries

Preface

This book provides an overview of the extensive research spanning more than four decades in the understanding and utilization of lithium (Li) metal anodes for rechargeable Li-metal batteries with a particular emphasis on the barriers, possible solutions, and potential applications in this important field. It may be served as a basic reference for readers interested in contributing to the further advancement of Li anodes and rechargeable Li-metal batteries.

We wish to express our sincere appreciation to all of the collaborators who have participated in our Li metal battery research. This work was supported by the Joint Center for Energy Storage Research (JCESR), an Energy Innovation Hub funded by the U.S. Department of Energy (DOE), Office of Science, Basic Energy Sciences and by the Advanced Batteries Materials Research (BMR) Program funded by the Assistant Secretary for Energy Efficiency and Renewable Energy (EERE), Office of Vehicle Technology of the DOE.

Richland, USA
Richland, USA
Research Triangle Park, USA

Ji-Guang Zhang
Wu Xu
Wesley A. Henderson

Contents

About the Authors

Dr. Ji-Guang (Jason) Zhang is a Laboratory Fellow of the Pacific Northwest National Laboratory (PNNL). He is the group leader for PNNL's efforts in energy storage for transportation applications and has 25 years of experience in the development of energy storage devices, including Li-ion batteries, Li-air batteries, Li-metal batteries, Li-S batteries, and thin-film solid-state batteries. He was the co-recipient of two R&D 100 awards, holds 18 patents (with another 17 patents pending), and has published more than 200 papers in refereed journals. He was named Thomson Reuters' Highly Cited Researchers-2015 in the Engineering category.

Dr. Wu Xu is a Senior Research Scientist in the Energy and Environment Directorate of Pacific Northwest National Laboratory. His main research interests include the development of electrolytes and electrode materials for various energy storage systems (such as lithium batteries, organic redox flow batteries and supercapacitors), the protection of lithium metal anode, and the investigation of electrode/electrolyte interfaces. He obtained his doctoral degree from the National University of Singapore in early 2000. He has had more than 120 papers published in peer-reviewed journals, six book chapters, and 25 U.S. patents granted with another 13 patents pending.

Dr. Wesley A. Henderson received his Ph.D. in Materials Science and Engineering (University of Minnesota) in 2002. He was a researcher at Lawrence Berkeley National Laboratory (1995) and Los Alamos National Laboratory (1996–1997), as well as an NSF International Research Fellow at ENEA, Advanced Energy Technologies Division, Rome, Italy. From 2004 to 2013, he was an Assistant Research Professor (Chemistry) at the U.S. Naval Academy and an Associate Professor (Chemical and Biomolecular Engineering) at North Carolina State University. He joined Pacific Northwest National Laboratory as a Senior Research Scientist in 2014 and joined the U.S. Army Research Office in 2016. His research expertise is liquid and solid electrolytes for energy storage/conversion applications including improved electrolyte characterization tools and methods and the correlation of molecular-level interactions with electrolyte properties and device performance.

Abbreviations

General Abbreviations

AEI	Anion exchange ionomer
AES	Auger electron spectroscopy
AFLB	Anode-free Li battery
AFM	Atomic force microscopy
AM-EFM	Amplitude-modulated electrostatic force microscopy
BF	Butyl formate
CE	Coulombic efficiency
CIP	Contact ion pairs
CV	Cyclic voltammetry
DOD	Depth of discharge
EDX	Energy dispersive X-ray spectroscopy
EFM	Electrostatic force microscopy
EQCM	Electrochemical quartz crystal microbalance
FOM	Figure of merit
FTIR	Fourier transform infrared spectroscopy
GC	Glass-ceramic
GC-MS	Gas chromatography–mass spectrometry
HAADF	High-angle annular dark field
HOMO	Highest occupied molecular orbital
IC	Ion chromatography
IL	Ionic liquid
LUMO	Lowest unoccupied molecular orbital
MAS	Magic angle spinning
MO	Molecular orbital
MRI	Magnetic resonance imaging
MS	Mass spectroscopy
NCA	Nickel-cobalt-aluminum oxide
N_{DMSO}	Solvation number of DMSO molecules
NMR	Nuclear magnetic resonance

OCP	Open circuit potential
PDOS	Projected density of states
PLE	Protected Li electrode
Raman	Raman spectroscopy
RPP	Reverse pulse plating
SEI	Solid electrolyte interphase
SEM	Scanning electron microscopy
SIMS	Secondary ion mass spectrometry
SPE	Solid polymer electrolyte
SPoM	Surface potential microscopy
SS	Stainless steel
SSE	Solid-state electrolyte
TEM	Transmission electron microscopy
TMAFM	Tapping-mode atomic force microscopy
TPD-MS	Temperature-programmed decomposition mass spectroscopy
UPD	Underpotential deposition
UPS	Underpotential stripping
XPS	X-ray photoelectron spectroscopy
XRD	X-ray diffraction

Chemical Abbreviations

2MeTHF	2-methyltetrahydrofuran
2MF	2-methylfuran
BETI	Bis(perfluoroethylsulfonyl)imide or $N(SO_2C_2F_5)_2^-$
BOB	Bis(oxalato)borate or $B[OC(=O)C(=O)O]_2^-$
DEC	Diethyl carbonate
DEE	1,2-diethoxyethane (or ethylene glycol diethyl ether)
DFOB	Difluoro(oxalato)borate or $BF_2[OC(=O)C(=O)O]^-$
DFT	Density functional theory
DMC	Dimethyl carbonate
DME	1,2-dimethoxyethane
DMSO	Dimethyl sulfoxide
DOL	1,3-dioxolane
EA	Ethyl acetate
EC	Ethylene carbonate
Et_2O	Diethyl ether
Et-G1	Ethylene glycol diethyl ether (i.e. 1,2-diethoxyethane)
FEC	Fluoroethylene carbonate
FSI	Bis(fluorosulfonyl)imide or $N(SO_2F)_2^-$
G2	Diglyme or diethylene glycol dimethyl ether
G3	Triglyme or triethylene glycol dimethyl ether
G4	Tetraglyme or tetraethylene glycol dimethyl ether
LATP	Lisicon-type $Li_{1+x}AL_xTi_{2-x}(PO_4)_3$
LiBETI	Lithium bis(perfluoroethylsulfonyl)imide

LiBOB	Lithium bis(oxalate)borate
LiDFOB	Lithium difluoro(oxalato)borate
LiFSI	Lithium bis(fluorosulfonyl)imide
LiTFSA	Lithium bis(trifluoromethanesulfonyl)amide or $LiN(SO_2CF_3)_2$
LiTFSI	Lithium bis(trifluoromethanesulfonyl)amide or $LiN(SO_2CF_3)_2$
MA	Methyl acetate
MF	Methyl formate
PC	Propylene carbonate
PEO	Poly(ethylene oxide)
PP	Polypropylene
PTFE	Polytetrafluoroethylene
TEGDME	Tetra(ethylene glycol) dimethyl ether
TFSA	Bis(trifluoromethanesulfonyl)amide or $N(SO_2CF_3)_2^-$
TFSI	Bis(trifluoromethanesulfonyl)imide or $N(SO_2CF_3)_2^-$
THF	Tetrahydrofuran
THP	Tetrahydropyran
VC	Vinylene carbonate
VEC	Vinylethylene carbonate

Chapter 1
Introduction

Lithium (Li) metal is an ideal anode material for rechargeable batteries due to its extremely high theoretical specific capacity (3860 mAh g^{-1}), the lowest negative electrochemical potential (−3.040 V versus standard hydrogen electrode), and low density (0.534 g cm^{-3}); thus rechargeable Li metal batteries have been investigated extensively during the last 40 years (Whittingham 2012; Aurbach and Cohen 1996; Xu et al. 2014). Unfortunately, rechargeable batteries based on Li metal anode have not yet been commercialized. There are two main barriers to the development of Li metal batteries. One is the growth of Li dendrites during repeated charge/discharge processes, and the other is the low Coulombic efficiency (CE) of these processes. These two barriers lead to two critical problems for Li metal anode: internal short circuits caused by dendrites—a safety hazard—and short cycle life of the battery due to reactions between Li metal and electrolyte, consumption of electrolyte, formation of inactive Li, and continuous increase of cell impedance. Although low CE can be partially compensated by an excess amount of Li—for example, an excess amount of 300 % of Li was a common solution in the early development of Li metal batteries—the dendrite growth-related battery failure, sometimes dramatic failure that led to fire and other hazards, and the emergence of Li-ion batteries have largely diminished industry's efforts on the development of rechargeable Li metal batteries since the early 1990s. Figure 1.1a, b show the schematic diagram of a typical Li-ion battery and a Li metal battery, respectively. In Li-ion batteries, graphite has been widely used as the anode material because Li ions can be intercalated into its layered structure so dendrite growth can be largely prevented. In Li metal batteries, the cathode shown in Fig. 1.1b can be replaced by sulfur (for a Li–S battery) or air electrode (for a Li-air battery). Figure 1.1c shows the typical morphology of a Li dendrite and the major problems associated with dendrites and low CE of the Li deposition/stripping processes.

With the urgent need for the "next generation" rechargeable batteries, such as Li–S batteries (Bruce et al. 2012; Ji et al. 2009), Li-air batteries (Girishkumar et al. 2010; Lee et al. 2011), as well as rechargeable Li metal batteries that use other Li

© Springer International Publishing Switzerland 2017
J.-G. Zhang et al., *Lithium Metal Anodes and Rechargeable
Lithium Metal Batteries*, Springer Series in Materials Science 249,
DOI 10.1007/978-3-319-44054-5_1

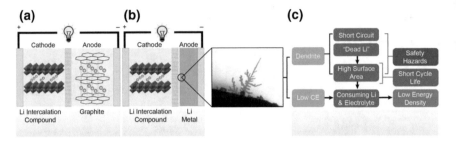

Fig. 1.1 Schematic diagram of **a** Li-ion batteries; **b** Li metal batteries; **c** The typical morphology of Li dendrite and main problems related to dendrites and low CE. Morphology picture in **c** is reproduced with permission, Copyright 1976, J. Crystal Growth (Chianelli 1976)

intercalation or Li conversion compounds as the cathode, the use of Li metal anodes has attracted significant interests in recent years. Over the last 40 years, Li dendrite formation has been widely analyzed (Aurbach and Cohen 1996; Chianelli 1976; Aurbach et al. 2002; Gireaud et al. 2006; Chandrashekar et al. 2012) and simulated (Monroe and Newman 2005; Tang et al. 2009; Yamaki 1998). Most approaches to dendrite prevention focus on improving the stability and uniformity of the solid electrolyte interphase (SEI) layer on Li surface by adjusting electrolyte components and optimizing SEI formation additives (Aurbach et al. 2002; Gireaud et al. 2006; Ota et al. 2004; Shiraishi et al. 1999; Lee et al. 2007). However, because metallic Li is thermodynamically unstable with organic solvents [as indicated by Aurbach et al. (2002)], it is very difficult to achieve sufficient passivation with a Li electrode in liquid solutions. As an alternative, various mechanical barriers, either ex situ coated polymer layers or inorganic solid-state blocking layers with high shear modulus, have been proposed to block dendrite penetration (Monroe and Newman 2005; Armand et al. 1981; Li et al. 2002; Balsara et al. 2008, 2009; Visco et al. 2004). These approaches rely on a strong mechanical barrier to prevent Li dendrite penetration through separator, but do not change the fundamental, self-amplifying behavior of the dendrite growth. In other words, these methods do not prevent Li dendrites from growing during long-term cycling, or barely improve the CE of Li deposition/stripping, thus they are not suitable for practical rechargeable Li batteries.

Although some factors that can suppress Li dendrite growth may also lead to high CE of Li cycling, in many experimental studies, they are not always directly correlated with each other. Overall, high CE is a more fundamental criterion required for stable cycling of a Li metal anode and the related Li metal batteries. To have a high CE, side reactions between native or freshly deposited Li and electrolyte have to be minimized. These side reactions are proportional to the chemical and electrochemical activity of native Li when it is in direct contact with the surrounding electrolyte. They are also proportional to the surface area of deposited Li. Therefore, a high CE of Li cycling is usually a direct result of low reactivity between freshly deposited Li and electrolyte as well as a low surface area of the deposited Li. On the one hand, a dendritic Li deposition always has a high surface area. This means that the high CE of Li deposition/stripping is always related to a low surface area Li deposition and a suppressed Li dendrite growth. A stable CE value during long-term cycling also means that an SEI layer formed during Li deposition is relatively stable and very minimal formation of new SEI layers occurs during each cycle. On the other hand, some electrolytes can lead to dendrite-free Li deposition, but exhibit a CE of only less than 80 % (Qian et al. 2015; Ding et al. 2013). This phenomenon often is related to a highly aligned nanoarray of Li structure that has no Li extrusion but still exhibits a high surface area, and freshly deposited Li is highly reactive with the surrounding electrolyte during the cycling process. Therefore, the enhancement of CE is a more fundamental factor controlling long-term, stable cycling of a Li metal anode.

In Chap. 2 of this book, we will first review various instruments/tools that are critical for the characterization of Li dendrite growth/stripping processes and analysis on the composition of the surface films formed during Li deposition, then we will review the general models of the dendrite growth processes. The effect of the SEI layer on the modeling of Li dendrite growth will also be discussed, which has often been neglected in the literature. In Chap. 3, various factors that affect CE of Li cycling and dendrite growth will be discussed, together with an emphasis on the enhancement of CE. This is partially due to the fact that almost all literature articles report the CE and Li deposition morphology together and a separate description will lead to significant duplications. Chapter 4 of the book will discuss the specific application of Li metal anodes in several key rechargeable Li metal batteries, including Li-air batteries, Li–S batteries, and Li metal batteries using other Li intercalation and Li conversion compounds as cathodes. Finally, a perspective on the future development and application of Li metal batteries will be discussed in Chap. 5.

References

Armand MB, Duclot MJ, Rigaud P (1981) Polymer solid electrolytes: stability domain. Solid State Ionics 3–4:429–430. doi:http://dx.doi.org/10.1016/0167-2738(81)90126-0

Aurbach D, Cohen Y (1996) The application of atomic force microscopy for the study of Li deposition processes. J Electrochem Soc 143(11):3525–3532

Aurbach D, Zinigrad E, Cohen Y, Teller H (2002) A short review of failure mechanisms of lithium metal and lithiated graphite anodes in liquid electrolyte solutions. Solid State Ionics 148:405–416

Balsara N, Panday A, Mullin SA (2008) Polymer electrolytes for high energy density lithium batteries. http://www1.eere.energy.gov/vehiclesandfuels/pdfs/merit_review_2008/exploratory_battery/merit08_balsara.pdf. Accessed 12 Jan 2016

Balsara NP, Singh M, Eitouni HB, Gomez ED (2009) High elastic modulus polymer electrolytes. US Patent Appl No 0263725 A1

Bruce PG, Freunberger SA, Hardwick LJ, Tarascon J-M (2012) Li-O$_2$ and Li-S batteries with high energy storage. Nat Mater 11(1):19–29

Chandrashekar S, Trease NM, Chang HJ, Du L-S, Grey CP, Jerschow A (2012) 7Li MRI of Li
    batteries reveals location of microstructural lithium. Nat Mater 11(4):311–315. doi:http://www.
    nature.com/nmat/journal/v11/n4/abs/nmat3246.html#supplementary-information
Chianelli RR (1976) Microscopic studies of transition metal chalcogenides. J Cryst Growth 34
    (2):239–244
Ding F, Xu W, Graff GL, Zhang J, Sushko ML, Chen X, Shao Y, Engelhard MH, Nie Z, Xiao J,
    Liu X, Sushko PV, Liu J, Zhang J-G (2013) Dendrite-free lithium deposition via self-healing
    electrostatic shield mechanism. J Am Chem Soc 135(11):4450–4456. doi:10.1021/ja312241y
Gireaud L, Grugeon S, Laruelle S, Yrieix B, Tarascon JM (2006) Lithium metal stripping/plating
    mechanisms studies: a metallurgical approach. Electrochem Commun 8(10):1639–1649.
    doi:10.1016/j.elecom.2006.07.037
Girishkumar G, McCloskey B, Luntz AC, Swanson S, Wilcke W (2010) Lithium—air battery:
    promise and challenges. J Phys Chem Lett 1(14):2193–2203
Ji XL, Lee KT, Nazar LF (2009) A highly ordered nanostructured carbon-sulphur cathode for
    lithium-sulphur batteries. Nat Mater 8:500–506
Lee YM, Seo JE, Lee Y-G, Lee SH, Cho KY, Park J-K (2007) Effects of triacetoxyvinylsilane as
    SEI layer additive on electrochemical performance of lithium metal secondary battery.
    Electrochem Solid-State Lett 10(9):A216–A219
Lee J-S, Kim ST, Cao R, Choi N-S, Liu M, Lee KT, Cho J (2011) Metal-air batteries with high
    energy density: Li–air versus Zn–air. Adv Energy Mater 1(1):34–50
Li Y, Fedkiw PS, Khan SA (2002) Lithium/V6O13 cells using silica nanoparticle-based composite
    electrolyte. Electrochimica Acta 47(24):3853–3861. doi:http://dx.doi.org/10.1016/S0013-4686
    (02)00326-2
Monroe C, Newman J (2005) The impact of elastic deformation on deposition kinetics at
    lithium/polymer interfaces. J Electrochem Soc 152(2):A396–A404. doi:10.1149/1.1850854
Ota H, Shima K, Ue M, Yamaki J-I (2004) Effect of vinylene carbonate as additive to electrolyte
    for lithium metal anode. Electrochim Acta 49(4):565–572. doi:10.1016/j.electacta.2003.09.010
Qian J, Xu W, Bhattacharya P, Engelhard M, Henderson WA, Zhang Y, Zhang J-G (2015)
    Dendrite-free Li deposition using trace-amounts of water as an electrolyte additive. Nano
    Energy 15:135–144. doi:10.1016/j.nanoen.2015.04.009
Shiraishi S, Kanamura K, Takehara Z (1999) Surface condition changes in lithium metal deposited
    in nonaqueous electrolyte containing HF by dissolution-deposition cycles. J Electrochem Soc
    146(5):1633–1639. doi:10.1149/1.1391818
Tang M, Albertus P, Newman J (2009). J Electro-chem Soc 156:A390–A399
Visco SJ, Nimon E, De Jonghe LC, Katz B, Chu MY (2004) Lithium fuel cells. In: Proceedings of
    the 12th international meeting on lithium batteries, Nara, Japan, June 27–July 2, 2004. The
    Electrochemical Society, Pennington, NJ
Whittingham MS (2012) History, evolution, and future status of energy storage. Proc IEE 100
    (Special Centennial Issue):1518–1534. doi:10.1109/JPROC.2012.2190170
Xu W, Wang J, Ding F, Chen X, Nasybulin E, Zhang Y, Zhang J-G (2014) Lithium metal anodes
    for rechargeable batteries. Energy Environ Sci 7(2):513–537. doi:10.1039/c3ee40795k
Yamaki J-I (1998) J Power Sources 74:219–227

Chapter 2
Characterization and Modeling of Lithium Dendrite Growth

Dendrites are a common occurrence when electrodepositing metals. Although the term "dendrite" is prevalent throughout the scientific literature when referencing Li deposition, such structures are atypical (i.e., in general, the plated Li morphology does not consist of regular branched, tree-like structures). Instead, Li tends to plate from solution as particles/nodules or whiskers/needles/filaments, which can aggregate into more complex constructs. Due to its ubiquitous usage for Li electrochemistry, however, the term "dendrite" will be used throughout the text in this book to refer to the latter structures. It is also notable that the literature that addresses the theory of Li electrodeposition has focused largely on the numerous determinant factors that influence classical dendritic metal plating, but the evolution of the plated Li structural characteristics is in actuality dictated to a great extent by the concurrent reactions between the reactive Li and electrolyte components (i.e., SEI formation).

2.1 Characterization of Lithium Dendrite Growth

As discussed in Chap. 1, the Li dendrite growth during the Li deposition process is a critical issue for the battery safety. Extensive efforts have been made to characterize and analyze the formation and growth processes of Li dendrites in order to reveal the mechanisms of dendrite formation and growth processes and then find the approaches to suppress or prevent the dendrite formation. In the last four decades, many different characterization methods have been used to study Li electrodes and the dendrite formation, including scanning electron microscopy (SEM) (Aurbach et al. 1998; Dollé et al. 2002), optical microscopy (Howlett et al. 2003; Nishikawa et al. 2010; Arakawa et al. 1993), atomic force microscopy (AFM) (Aurbach and Cohen 1997), transmission electron microscopy (TEM) (Liu et al. 2011; Ghassemi et al. 2011), nuclear magnetic resonance (NMR) (Chang et al. 2015), Fourier transform infrared spectroscopy (FTIR) (Morigaki 2002; Aurbach et al. 1987, 1995,

© Springer International Publishing Switzerland 2017
J.-G. Zhang et al., *Lithium Metal Anodes and Rechargeable Lithium Metal Batteries*, Springer Series in Materials Science 249, DOI 10.1007/978-3-319-44054-5_2

1997; Kanamura et al. 1997), and x-ray photoelectron spectroscopy (XPS) (Ota et al. 2004a; Aurbach et al. 1993; Kanamura et al. 1995b). Both morphology and chemical composition of the deposited Li metal films have been extensively investigated. The main approaches used in the characterization of electrochemically deposited Li metal films are briefly introduced in this chapter.

2.1.1 Characterization of Surface Morphologies

2.1.1.1 SEM

Among various observation methods, SEM is the most common and useful technique to examine Li electrode surface morphologies. Morphology study focuses on variation of the Li surface and formation of Li dendrites. Since the 1970s, this method has been used to study Li film growth (Dey 1976) both ex situ and in situ during dendrite growing/stripping processes (Ding et al. 2013, 2014; Zhamu et al. 2012; Ryou et al. 2012; Stark et al. 2011; Aurbach et al. 1989, 1990a, 1994b; Besenhard et al. 1987; Kanamura et al. 1995a; Yamaki et al. 1998; Shiraishi et al. 1999; Ota et al. 2004a; Gireaud et al. 2006; López et al. 2009; Nazri and Muller 1985b; Yoshimatsu et al. 1988; Choi et al. 2011; Lee et al. 2006; Yoon et al. 2008; Li et al. 2014a; Bieker et al. 2015). With its high resolution images, SEM is a powerful technique to analyze the surface change of Li metal during the deposition and stripping cycles. The effects of solvents (Aurbach et al. 1990a, b; Besenhard et al. 1987; Ota et al. 2004a; Bieker et al. 2015; Naoi et al. 1999), salts (Aurbach et al. 1994; Kanamura et al. 1995a; Yang et al. 2008), additives, (Shiraishi et al. 1999; Fujieda et al. 1994; Osaka et al. 1997b; Ota et al. 2004b; Lee et al. 2007), and other treatments (Stark et al. 2011; Thompson et al. 2011; Stark et al. 2013) have been directly discovered using SEM images. With the help of SEM observations, the correlation between the surface chemistry and morphology of Li electrodes was built (Aurbach et al. 1994), and the morphological transitions on Li metal anodes during cycling were examined (López et al. 2009). In the ex situ SEM studies, the Li samples have been transferred into the SEM instrument chamber in an inert atmosphere to avoid reaction of Li metal with ambient air as described by Kohl and coworkers (Stark et al. 2011). In the in situ SEM observation of Li dendrite formation, Orsini et al. (1998, 1999) first reported using in situ SEM to observe the cross section of plastic rechargeable Li batteries using solid polymer electrolytes. They observed the accumulation of mossy Li and growth of dendritic Li at the Li/polymer electrolyte (Fig. 2.1), which was the origin of rapid interface deterioration and capacity fading.

Neudecker et al. (2000) used in situ SEM to observe variation in the behavior of the anode current collector and the overlayer during Li deposition and stripping in an in situ built solid-sate thin-film battery. The cross-sectional SEM image of the battery is shown in Fig. 2.2a. During the initial charge of the battery at 4.2 V, Li was plated between the anode current collector (i.e., Cu) and the solid-state

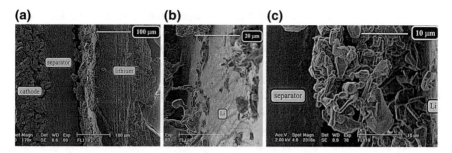

Fig. 2.1 **a** Cross section of a Li battery after one charge at 1C; **b** surface of the Li anode of a Li battery after one charge at 1C; **c** Li deposit on the Li surface after one charge at 1C. The 1C rate was 2.2 mA cm^{-2}. Reproduced with permission—Copyright 1998, Elsevier (Orsini et al. 1998)

electrolyte (i.e., lithium phosphorous oxynitride or LiPON), and lifted the Cu/LiPON overlayer but did not significantly penetrate the Cu film (Fig. 2.2b). During discharge to 3.0 V, the plated Li was stripped from the Cu substrate but the inactive Li prevented the exposed edge of the Cu/LiPON overlayer from completely settling onto the LiPON electrolyte again (Fig. 2.2c). Recently, Sagane et al. (2013) also reported use of in situ SEM to observe Li plating and stripping reactions at the LiPON/Cu interface. They found that nucleation reactions are the rate-determining step during the Li plating process, while the Li$^+$ cation diffusivity governs the stripping process.

Nagao et al. (2013) also used in situ SEM to study the Li deposition and dissolution mechanism in a bulk-type solid-state cell with a Li$_2$S-P$_2$S$_5$ solid-state electrolyte (SSE). They reported that at a deposition current density over

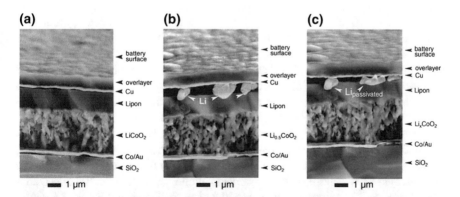

Fig. 2.2 In situ SEM micrographs showing cross-sectional views of a Li-free thin-film battery with an in situ plated Li anode: **a** 1 μm LiPON overlayer/2000 Å Cu anode current collector/1.5 μm LiPON electrolyte/3.0 μm LiCoO$_2$ cathode on quartz glass substrate prior to the initial charge, **b** same cell but at the end of the initial charge to 4.2 V, and **c** same cell but at the end of the first discharge to 3.0 V. The markers indicate a length of 1 μm. Reproduced with permission—Copyright 2000, The electrochemical society (Neudecker et al. 2000)

1 mA cm^{-2}, the local Li deposition triggered large cracks, resulting in a decrease in the reversibility of Li deposition and dissolution. On the other hand, at a low current density of 0.01 mA cm^{-2}, Li deposition was homogeneous thus greatly reducing the occurrence of unfavorable cracks, which enables reversible deposition and dissolution. These results suggest that homogeneous Li deposition on the surface of the SSE and suppression of the growth of Li metal along the grain boundaries inside the SSE are the keys to achieve the repetitive Li deposition and stripping without deterioration of the SSE. Clearly, SEM (either ex situ or in situ) is a very useful technique to investigate morphology variations of Li electrodes during deposition and stripping in open cells (Li et al. 2014a). However, because of the ultrahigh vacuum condition in SEM tests, most in situ SEM investigations of Li deposition/stripping processes are performed on batteries using solid polymer or inorganic SSE. It is still difficult to conduct an in situ observation of Li formation and growth in conventional liquid electrolytes, which is more relevant to most of practical applications.

2.1.1.2   Optical Microscopy

Optical microscopy is another way to observe the Li dendrites and it has been widely used as an in situ method to observe and record the Li dendrite growing/stripping processes under working conditions close to those of practical applications (Osaka et al. 1997a). Although the resolution of optical microscopy is not as high as that of SEM, it still could easily and instantaneously distinguish the surface change and dendrite formation. It is an intuitional observation on Li electrodes and very helpful to understand a continuous dendrite growing/stripping process. With digital recording devices, the dendrite formation process can be recorded as a video. Therefore, the optical microscopy technique has been widely used to analyze Li electrodes in situ. However, a special optical cell is needed for in situ optical study of Li dendritic growth. Usually, such a cell is homemade as described by Brissot et al. (1998). The optical cell could work as an airtight electrochemical cell, and the Li electrode surface could be observed by optical microscopy. In the in situ observation, optical microscopy is more often used to study the cross section of a Li electrode to observe the dendrite growth perpendicular to the surface of the Li electrode (Stark et al. 2013; Brissot et al. 1998, 1999b; Sano et al. 2011; Howlett et al. 2003; Hernandez-Maya et al. 2015). Figure 2.3 shows typical dendrites formed at different current densities during in situ optical microscopy study. At low current density (0.2 mA cm^{-2}), needle-like and particle-like dendrites are observed, while at higher current densities (≥ 0.7 mA cm^{-2}), dendrites have a tree-like or bush-like shape. Figure 2.4 shows the evolution of Li dendrites observed in the inter-electrode space while the cell is being polarized (at 0.05 mA cm^{-2}). The needle-like dendrite grows on the negative electrode and finally contacts the positive electrode, causing short circuit of the cell.

Fig. 2.3 Typical dendrites obtained with different current densities: **a** $J = 0.2$ mA cm^{-2} (needle-like dendrites), **b** $J = 0.7$ mA cm^{-2} (tree-like dendrites), **c** $J = 1.3$ mA cm^{-2} (bush-like dendrites). Reproduced with permission—Copyright 1998, Elsevier (Brissot et al. 1998)

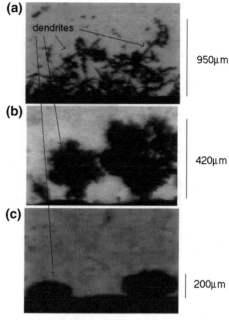

Fig. 2.4 Time variation of the dendrites observed in the inter-electrode space while polarizing the cell with $J = 0.05$ mA cm^{-2}. Dendrites are seen to be needle-like. It shows that the dendrite grows on negative electrode with time and finally contacts the positive electrode, causing short circuit of the cell. Reproduced with permission —Copyright 1999, Elsevier (Brissot et al. 1999b)

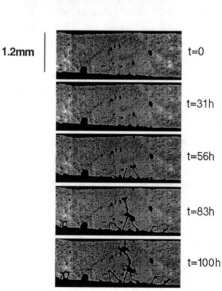

2.1.1.3 AFM

AFM is another useful technique to investigate Li electrode morphology. The resolution of AFM is much better than that of optical microscopy. At the same time, AFM can give a three-dimensional (3D) morphology image that is difficult to get

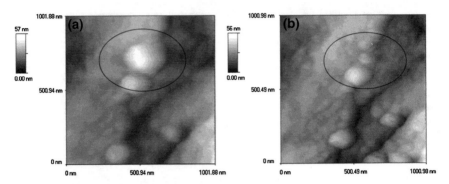

Fig. 2.5 AFM images (1 × 1 μm) of a Li electrode in a 0.5 M LiAsF₆/PC solution. **a** an image obtained after Li deposition, 0.41 C cm⁻². New Li deposits are marked by a circle. **b** an image of the same area after consecutive Li dissolution (0.41 C cm⁻²). The same area marked in **a** is also circled here. Reproduced with permission—Copyright 2000, American Chemical Society (Cohen et al. 2000)

from SEM or optical microscopy. In 1996, Aurbach and Cohen first used AFM to study the Li deposition processes in nonaqueous electrolyte systems (Aurbach and Cohen 1996). In that work, the basic electrochemical cell was modified to hold the highly sensitive electrodes and electrolyte solution and to isolate them from atmospheric contaminants. They found that the AFM scanning is not destructive and does not change the morphology on the surface. After that, more work (Morigaki and Ohta 1998; Aurbach et al. 1999; Cohen et al. 2000; Morigaki 2002; Mogi et al. 2002a, b, c) using in situ AFM was conducted. With the special 3D morphology of AFM images, the swelling and shrinking of Li surfaces during Li deposition and stripping processes have been discovered (Morigaki 2002). Figure 2.5 shows AFM images of a Li surface film deposited in a 0.5 M LiAsF₆/ propylene carbonate (PC) electrolyte, where (a) a bump after Li deposition and (b) shrinkage after consecutive Li stripping are clearly seen (Morigaki 2002). Moreover, investigation by AFM has revealed that the structure of the Li surface consists of grain boundaries, ridge lines, and flat areas (Morigaki and Ohta 1998), which cannot be proven by other morphology test methods including SEM and optical microscopy. Based on these findings, the breakdown and reparation of the SEI films on Li electrodes during Li deposition/stripping cycles have been proposed (Fig. 2.6) and probed (Aurbach and Cohen 1996; Cohen et al. 2000).

Several modified AFM methods have also been used in the characterization of Li film deposition. Shiraishi et al. reported using in situ fluid tapping-mode AFM (TMAFM) coupled with an electrochemical quartz crystal microbalance to investigate the electrochemical stripping behavior of Li metal in nonaqueous electrolytes containing a small amount of HF (Shiraishi and Kanamura 1998), and also using TMAFM with surface potential microscopy (SPoM) to study the surface condition of the electrodeposited Li on a Ni substrate (Shiraishi et al. 2001). Recently, Zhang et al. (2014) used amplitude-modulated electrostatic force microscopy (AM-EFM) (a special type of AFM) to study the surfaces of deposited Li films. The Li films

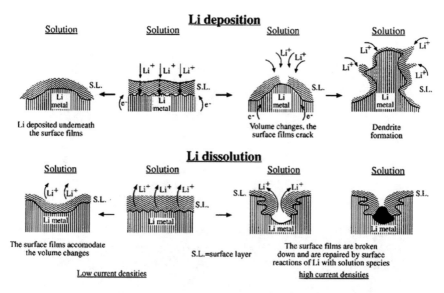

Fig. 2.6 A description of the morphology and failure mechanisms of Li electrodes during Li deposition and dissolution illustrating selected phenomena: the beginning of dendrite formation (*top*) and nonuniform Li dissolution accompanied by breakdown and reparation of the surface films (*bottom*). Reproduced with permission—Copyright 2000, American Chemical Society (Cohen et al. 2000)

were electrodeposited for 15 h on Cu foils in electrolytes of 1 M LiPF$_6$-PC without and with 0.05 M CsPF$_6$ additive. Li dendrites were formed in the control electrolyte (i.e., without Cs$^+$ additive) and a smooth Li film was obtained in the Cs$^+$ containing electrolyte. As shown in Fig. 2.7, the EFM images recorded at a probe voltage of -1 V for the Li film formed in the control electrolyte exhibit wide color variations or a strong contrast, which indicates a large fluctuation and nonuniform distribution of electric field across the detected Li surface. In comparison, the EFM image for the Li film formed in the Cs$^+$-containing electrolyte shows narrow color variations or relatively small contrast, indicating a uniform distribution of the electric field across the Li surface and is consistent with a smooth Li film.

2.1.1.4 TEM

Due to the success of the application of optical microscopy in the observation of morphologies of deposited Li films in microscale, another electron microscopy, i.e., in situ TEM, was recently used to observe in real time the formation of Li fibers or Li dendrite growth at nanoscale (Liu et al. 2011; Ghassemi et al. 2011). Huang and coworkers first reported the direct observation of Li fiber growth on different nanowire anodes (such as silicon or tin oxide) during in situ charging of nanoscale Li-ion batteries inside a TEM (Huang et al. 2010). Li fibers of up to 35 μm long

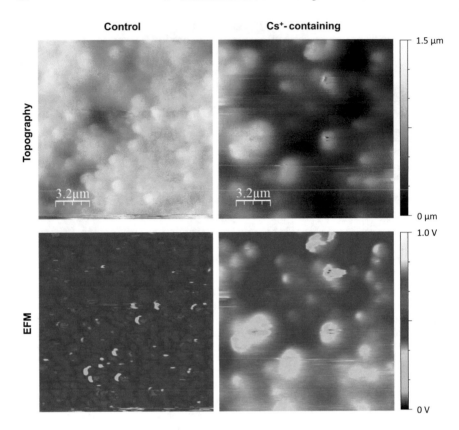

Fig. 2.7 Topography and EFM images (recorded at a probe voltage of −1 V and frequency of 22 kHz) of a dried Li film after deposition from the control electrolyte without Cs⁺-additive (i.e., 1.0 M LiPF₆/PC) and the Cs⁺-containing electrolyte (i.e., 1.0 M LiPF₆-PC with 0.05 M CsPF₆). The EFM images show the distribution of electric field across the Li surface. The *wide color* variations in the EFM images obtained in the control electrolyte indicate a nonuniform distribution, while the *narrow color* variations exhibited in the images obtained in Cs⁺-containing electrolyte indicate a more uniform distribution of electric field across the detected Li surface. Reproduced with permission—Copyright 2015, American Chemical Society (Zhang et al. 2014)

grew on nanowire tips along the nanowire axis in an ionic liquid (IL)-based electrolyte. After that, Yassar and coworkers also used this in situ TEM technology to study the growth of Li dendrites (Ghassemi et al. 2011). They reported clear observation of nucleation of Li⁺ cations at the anode/electrolyte interface and then growth of Li fibers or Li dendrites on the anode surface in a nanoscale Li-ion battery (Fig. 2.8). In situ TEM is a very promising method to observe and study Li dendrite growth in situ during the continuous charging/discharging processes of a battery, especially at the nanoscale.

Although in situ TEM images have been used to reveal the formation and growth of Li dendrites during the continuous charging/discharging processes of a battery, an IL

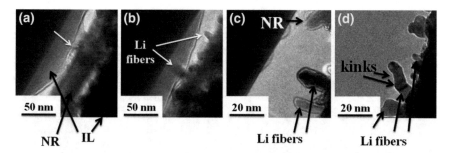

Fig. 2.8 a *Black arrows* indicate an individual silicon nanorod surrounded by ionic liquid. **b** *Arrows* indicate the formation of Li islands on the nanorod. **c** Represents the growth of Li fibers. **d** The formation of kinks and growth of Li fibers are marked by *black arrows*. Reproduced with permission—Copyright 2011, American Institute of Physics (Ghassemi et al. 2011)

electrolyte or an SSE has to be used in most in situ TEM studies because the high vapor pressure of a practical liquid electrolyte is not compatible with the high vacuum required by a conventional TEM system. It is well known that an SEI layer formed on the surface of a Li film or dendrite is critical for Li deposition/stripping. However, the SEI formed in IL or SSE is greatly different from those formed in the conventional electrolytes used in Li metal batteries. Therefore, the interaction between Li dendrites and a practical liquid electrolyte still cannot be observed in these in situ TEM studies. Very recently, with the development of liquid cells for in situ TEM techniques, a true operando TEM investigation on Li dendrite growth has been performed in electrochemical cells with conventional liquid electrolytes for Li-ion batteries (Gu et al. 2013; Mehdi et al. 2014, 2015; Sacci et al. 2014). Sacci et al. (2014) reported the direct visualization of an initial dendritic SEI formation prior to Li deposition, and this dendritic morphology remained on the surface after Li dissolution during the in situ electrochemical TEM study, which used 1.2 M $LiPF_6$ in ethylene carbonate (EC)/ dimethyl carbonate (DMC) (3:7 by wt) as electrolyte. Mehdi et al. (2015) used in situ electrochemical scanning TEM (STEM) to study the initial stages of SEI formation and Li dendrite evolution at the anode/electrolyte interface during Li deposition/stripping processes in 1 M $LiPF_6$/PC electrolyte. The high-angle annular dark field (HAADF) STEM images of the anode/electrolyte interface during the first three charge-discharge cycles of this operando Li battery are shown in Fig. 2.9. The deposition and stripping of Li is clearly observed. Some electrochemically inactive or "dead" Li residues around the electrode after Li stripping for all the three cycles are present, which are no longer attached to the Pt electrode.

Presently, the liquid cell TEM gives relatively lower resolution images than the open cell (i.e., vacuum conditioned) TEM does due to the limitation of cell thickness required to hold liquid electrolyte. Therefore, future improvement is required on the liquid cell fabrication (including electrodes with different alloying performance and control of the thickness) along with use of faster imaging methods. Such improvements should enable the clear observations of the initial stages of different mechanisms to be quantified on the nanometer to atomic scale. When

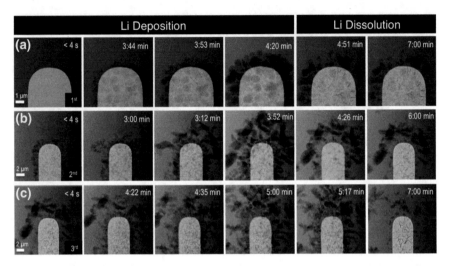

Fig. 2.9 HAADF images of Li deposition and dissolution at the interface between the Pt working electrode and the LiPF$_6$-PC electrolyte during the **a** first, **b** second, and **c** third charge/discharge cycles of the operando cell. The formation of the SEI layer (ring of contrast around the electrode), alloy formation due to Li$^+$ ion insertion, and the presence of "dead Li" detached from the electrode can all be seen in the images at the end of the cycle, thereby demonstrating the degree of irreversibility associated with this process. Reproduced with permission—Copyright 2015, American Chemical Society (Mehdi et al. 2015)

coupled with different electrolyte compositions (i.e., solvents, salts, and additives), the improved liquid cell TEM technology may provide critical insights into the complex interfacial reactions for future Li-based and other next-generation energy storage systems.

2.1.1.5 NMR

NMR is a powerful tool for detecting chemical bonds or atomic surroundings. Recently, Bhattacharyya and Grey et al. proposed using the difference between NMR signal intensities of bulk and porous Li to identify the Li dendrite growth (Bhattacharyya et al. 2010). They have successfully used this method as an in situ tool to quantitatively observe the formation of Li dendrites in different electrolytes. Chandrashekar et al. (2012) reported using ^{7}Li magnetic resonance imaging (MRI) technique to detect in situ the variation of Li electrode morphologies during charge and discharge processes of a symmetric Li metal cell. The ^{7}Li NMR spectra of the Li metal resonance before (pristine) and after applying a current (charged) indicated that the area of the spectrum in the charged state was 2.3 times larger than that in the pristine state. This increase could be attributed to the formation of

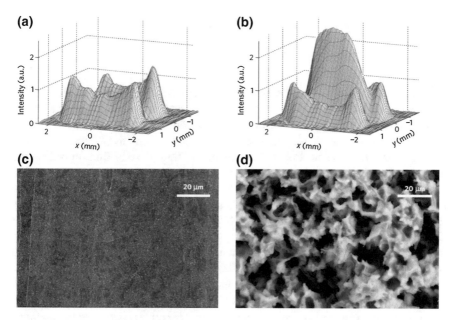

Fig. 2.10 ^{7}Li 2D MRI x-y images (frequency encoding in x and phase encoding in y) in the states of pristine (**a**) and after charging (**b**), and the related SEM images of Li anode in pristine (**c**) and charged (**d**) states. Reproduced with permission—Copyright 2012, Macmillan Publishers (Chandrashekar et al. 2012)

dendritic, mossy, and other microstructural metallic Li during charging. The two-dimensional ^{7}Li MRI images before and after the cell charging are depicted in Figs. 2.10a, b, where the cumulative signals were projected along the z direction which is perpendicular to the substrate. The MRI image of the charged battery revealed the negative electrode had a significant increase in signal of almost double, while the positive electrode showed a decrease in signal of about 23 % after charging. It indicated the location and change of microstructural Li morphology, which is consistent with findings from SEM images (Figs. 2.10c, d). Recently, Arai et al. (2015) used in situ solid-state ^{7}Li NMR to observe Li metal deposition during overcharge in Li-ion batteries. Hu et al. also used in situ NMR and computational modeling to investigate the role of Cs^+ additive (Hu et al. 2016). These works indicate that NMR not only can detect the morphology variation during the Li metal deposition process, but also can reveal the possible composition of SEI layers formed on the surfaces of Li films during the electrode plating process. By combining NMR and other characterization techniques, a more comprehensive understanding of the electrode plated Li films can be obtained.

2.1.2 Characterization Methods for Surface Chemistry

2.1.2.1 FTIR

The characterization methods discussed in the previous sections mainly provide information on the morphology variations of Li depositions. Several other methods have been used to analyze the chemical compositions of the surface films formed on the surface of deposited Li. The chemical composition of the Li surface film is strongly related to the electrolyte components. In turn, the SEI film formed on the surface of a Li deposition strongly affects the Li morphology and cycling performance of a Li metal battery (Aurbach et al. 1994). In this aspect, FTIR and XPS are widely used in this field to analyze the Li surface chemistry; FTIR is more suitable for detecting the organic components, while XPS gives more information about the inorganic components.

Since the 1980s, FTIR has been widely used to analyze the Li surface as a nondestructive method (Morigaki 2002; Aurbach et al. 1987, 1995, 1997; Kanamura et al. 1997). In the early years, FTIR was used as an ex situ method (Aurbach et al. 1987), but it was later developed as an in situ technique to analyze Li films during electrochemical processes (Morigaki 2002; Kanamura et al. 1997). FTIR has been used to investigate the influences of electrolyte solvents, salts, additives, and other contaminants on the Li surface. From the locations and strengths of the peaks in FTIR spectra, different chemical bonds or components on the Li electrode surface could be identified. An example is shown in Fig. 2.11 which shows FTIR spectra of Li electrodes prepared and stored for three days in EC-DMC solutions of 1 M $LiAsF_6$, $LiPF_6$, or $LiBF_4$ as indicated. A spectrum of a Li electrode prepared and stored in DMC containing 0.1 M CH_3OH is also presented for comparison (Aurbach et al. 1997).

2.1.2.2 XPS

It should be noted that although the FTIR technique is very useful to identify the surface components, it is limited in that it detects only the IR-active species and it cannot give the relative importance of each surface component and composition that affect the Li deposition morphology and battery performance. Therefore, other surface characterization methods and technologies besides FTIR are needed to get more detailed surface chemistry data on Li anodes. As mentioned above, XPS is another very useful tool to analyze the surface chemistry of Li electrodes; in particular, it gives more information about the elemental or inorganic components—data that FTIR cannot detect. Normally, as indicated from XPS and FTIR data by Aurbach et al. (1987), the major species in a Li surface film include Li_2O, LiOH, LiF, Li_2CO_3, lithium alkylcarbonate (RCOOLi), and hydrocarbon. Recently,

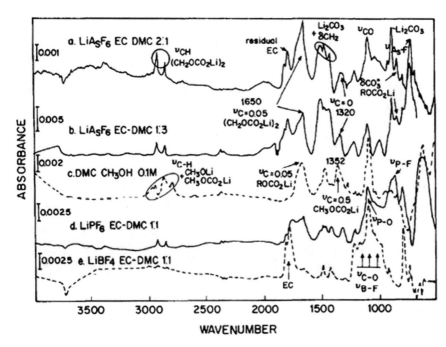

Fig. 2.11 FTIR spectra of Li electrodes prepared and stored for three days in EC-DMC solutions of 1 M LiAsF$_6$, LiPF$_6$, or LiBF$_4$ as indicated. A spectrum of a Li electrode prepared and stored in DMC containing 0.1 M CH$_3$OH is also presented for comparison. Reproduced with permission— Copyright 1997, Elsevier (Aurbach et al. 1997)

Wenzel et al. (2015) used the in situ XPS technique to study the stability of an SSE in contact with Li metal. The key concept was to use the internal Ar ion sputter gun in a standard lab-scale photoelectron spectrometer to deposit thin metal films (e.g., Li) on the sample surface and to study the reactions between metal and SSE by photoelectron spectroscopy directly after deposition (Fig. 2.12). This approach could give information on interfacial reactions and the interfacial kinetics, especially for the interface between the alkali metal and solid electrolyte in solid-state batteries.

With XPS analysis, the effects of different electrolyte solvents (Ota et al. 2004a; Aurbach et al. 1993; Kanamura et al. 1995b), lithium salts (Kanamura et al. 1995a; Fujieda et al. 1994), and additives (Shiraishi et al. 1999; Odziemkowski et al. 1992) on the Li surface chemistry have been investigated. Based on these data, especially with the vacuum etching technology of the XPS technique, not only the components but also the structural composition evolution of the SEI film can be revealed by XPS analysis (Ding et al. 2014; Kanamura et al. 1995a; Shiraishi et al. 1999; Zhang et al. 2014; Aurbach et al. 1993; Qian et al. 2015a, b).

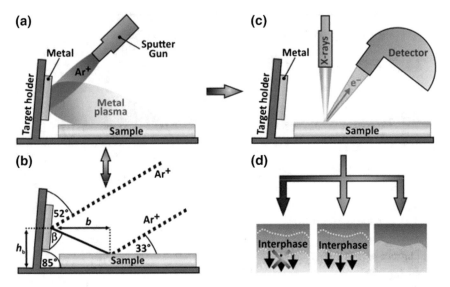

Fig. 2.12 The basic concept and setup of the in situ XPS technique. An argon ion beam is used to sputter lithium, gold or aluminum metal on the sample surface (**a**). In (**b**) the geometry is schematically shown. After deposition, the reaction products formed at the interface (**d**) are investigated using photoelectron spectroscopy, as shown in (**c**). Reproduced with permission—Copyright 2015, Elsevier (Wenzel et al. 2015)

2.1.3 Other Characterization Techniques

In addition to the methods discussed in the above sections, several other methods, including Raman spectroscopy (Raman) (Kominato et al. 1997), Auger electron spectroscopy (AES) (Ota et al. 2004a; Morigaki and Ohta 1998; Aurbach et al. 1993), and NMR (Kominato et al. 1997) have been used to analyze the surface chemistry of electroplated Li films. So far, attempts to use Raman spectroscopy to identify the surface films on the Li metal/electrolyte interphase have not been very successful. Only a few papers reported the Raman studies (Howlett et al. 2003; Rey et al. 1998a, b; Naudin et al. 2003). For example, Irish et al. used a Raman microprobe to study both in situ and ex situ the surface films formed on Li metal in contact with electrolytes of $LiAsF_6$/tetrahydrofuran (THF) and $LiAsF_6$/2MeTHF (Odziemkowski et al. 1992). The reaction products detected were mainly polytetrahydrofuran, some arsenolite (As_2O_3), and arsenious oxyfluorides $F_2As\text{-}O\text{-}AsF_2$. Raman technology might be expected to yield results similar to those of FTIR spectroscopy, but this technology is more complicated to use than FTIR (Odziemkowski et al. 1992). In addition, as indicated by Naudin et al. (2003), local heating of the samples under laser irradiation is unavoidable in Raman tests. The carbonate species on Li surface could be transformed into lithium acetylides of Li_2C_2 type, which gives a vibration peak of $C\equiv C$ at about 1845 cm^{-1}, thus giving a faulty result to the interpretation. Therefore, the destructive effect of Raman laser beam on Li surfaces limits its use in the analysis of Li surface films.

Aurbach et al. (1993) used AES to measure the Li surface after the Li was immersed in an electrolyte of 0.2 M $LiAsF_6$/1,2-dimethoxyethane (DME) for 15 min followed by pure DME rinsing. They found that the AES spectra were similar to those seen with XPS. Carbon and oxygen were detected at the Li surface. With sputtering, the intensity of the carbon Auger peak decreased while the oxygen peak increased when compared to their initial peaks. It was suggested that the surface films of Li treated in DME consisted of two layers, the upper layer being an alkoxide film (probably $LiOCH_3$) and the layer close to Li being a mixture of Li_2O and LiOH. Kominato et al. (1997) also used AES to detect the surface compounds of Li after it was immersed in three electrolytes of EC/DMC with $LiPF_6$, $LiClO_4$, or LiTFSI. Except for LiF found in the Li surface film from the $LiPF_6$-based electrolyte, all major components in the three Li surfaces were Li–O components indicating LiOH, Li_2O, or other lithium oxide compounds. Morigaki and Ohta used scanning AES to analyze the Li surface in 1 M $LiClO_4$/PC solution also used AES to detect the surface compounds of Li after it was immersed in three electrolytes of EC/DMC with $LiPF_6$, $LiClO_4$, or LiTFSI. Except for LiF found in the Li surface film from the $LiPF_6$-based electrolyte, all major components in the three Li surfaces were Li–O components indicating LiOH, Li_2O, or other lithium oxide compounds. Morigaki and Ohta (1998) used scanning AES to analyze the Li surface in 1 M $LiClO_4$/PC solution. Li_2CO_3, Li_2O, and LiOH were localized on the ridge lines and the grain boundaries of the Li surface. AES technology can provide some useful information about the Li surface components, but it is very close to that obtained from XPS analysis. In addition, the AES equipment is more difficult to access than XPS equipment, so AES analysis has been used less frequently in Li metal investigations.

NMR has also been used to study the Li surface chemistry. Ota et al. (2004a, c) used NMR technology (^1H, ^{13}C and 2D spectra) to analyze the surface components of deposited Li by dissolving the surface film in anhydrous dimethyl sulfoxide (DMSO)-d_6 and then recording the NMR spectra of the organic species in the DMSO-d_6 solution. They found that the organic surface layer on Li metal included lithium ethoxide, lithium ethylene dicarbonate, PEO, and lithium ethylene containing an oxyethylene unit. This is an indirect method to analyze the Li surface focusing on the dissolvable organic species. The obvious limitation of this technique is the inability to analyze the insoluble inorganic compounds formed on Li surfaces.

Nazri and Muller used secondary ion mass spectrometry (SIMS) to study the surface layer formed on electrochemically deposited Li on copper in a 1 M $LiClO_4$-PC electrolyte (Nazri and Muller 1985b). The obtained SIMS spectrum was complex and was difficult to interpret. Basically, the low mass range showed the fragments of PC, the salt, and water, while the high mass range indicated the presence of a polymeric material based on PC, a partially chlorinated hydrocarbon polymer, and their lithium adducts. The authors also applied the in situ x-ray diffraction (XRD) technique to the analysis of the formed Li surface films (Nazri and Muller 1985a, b, c). The presence of Li_2CO_3, Li_2O, and polymer compounds was also detected.

Temperature-programmed decomposition mass spectroscopy (TPD-MS) and gas chromatography–mass spectrometry (GC-MS) technology were also used by several groups to analyze the surface components of Li electrodes (Matsuda et al. 1995). Kominato et al. (Morigaki and Ohta 1998; Kominato et al. 1997) reported that the gases generated from Li films pretreated in EC-dimethyl carbonate (DMC) based electrolytes were mainly CH_4, H_2O, CO, CH_3OH, CO_2, and ethylene oxide. N_2 was also detected if LiTFSI was used as the electrolyte salt. The GC-MS detected the same gas components. This indicated that the detected gases were generated from the organic Li compounds that were the reaction products of Li and solvents (EC and DMC) and included lithium ethylene dicarbonate and lithium methylcarbonate. Ota et al. (2004a, c) also used GC-MS to investigate the Li surface compounds generated in EC/THF electrolytes. C_2H_4, CO_2, and C_2H_6 were detected and were mainly from the reductive components of EC. During TPD-MS and GC-MS measurements, the Li electrodes with the detected surface films need to be heated to give off the gases to be tested. The data from these MS measurements could provide more information on the Li surface film chemistry and support the results of other film measurements, such as FTIR and XPS.

Ota et al. (2004a, c) used ion chromatography (IC) to quantitatively analyze the Li surface films. The Li films were first dissolved in high-purity water and then tested by an IC instrument. By analyzing the contents of F^-, CO_3^{2-}, and Li^+ ions, quantitative information about the Li surface films could be obtained. They found that the Li surface film in EC-based electrolytes consisted mainly of lithium alkyl carbonate, and LiF content in the films formed in an electrolyte containing lithium imide salt was lower than those formed in the electrolytes containing $LiPF_6$ salt.

The in situ scanning vibrating electrode technique has also been used to map the surface electric field of Li electrodes (Matsuda et al. 1995; Ishikawa et al. 1997). The surface electric field on a Li electrode is based on its morphology and composition uniformity. So this technology reflects not only the surface morphology, but also the chemical composition uniformity of the Li surface. However, because the scanning step of this technology is not small enough, the definition obtained using this technology is not satisfactory.

In a recent effort, Harry et al. (2014, 2015) used synchrotron hard x-ray microtomography to investigate the failure caused by dendrite growth in high-energy density, rechargeable batteries with Li metal anodes. When a symmetric Li|polymer electrolyte|Li cell was cycled at 90 °C, they found that the bulk of the dendritic structure lay within the electrode, underneath the polymer/electrode interface, during the early stage of dendrite development. Furthermore, they observed crystalline impurities, present in the uncycled Li anodes, at the base of the subsurface dendritic structures. The portion of the dendrite protruding into the electrolyte increases on cycling until it spans the electrolyte thickness, causing a short circuit. Contrary to conventional wisdom, it seems that preventing dendrite formation in polymer electrolytes depends on inhibiting the formation of subsurface structures in the Li electrode. These results demonstrate that x-ray microtomography is another powerful tool to provide a clear failure mechanism in Li metal batteries.

In summary, characterization of morphologies and surface components of electroplated Li anodes is a complicated task. SEM and FTIR/XPS are the most common methods used to investigate the surface morphology and chemistry, respectively, of the electrodeposited Li films. Many other methods discussed in this chapter also provide valuable information on Li films. However, no single technique is enough to provide comprehensive understanding of the studied Li films, especially for the surface reaction products or SEI layer formed on a Li film. Therefore, a combination of multiple characterization and analysis methods is required to have a good understanding of the properties of electrodeposited Li films.

2.2 Effect of SEI Layer on Lithium Dendrite Growth

Various models proposed in the literature have provided important guidance on the nucleation and growth of metal dendrites, especially for metals that do not react with electrolyte in a significant way. However, Li is thermodynamically unstable with any organic solvent and the two react instantaneously to form an SEI. This SEI layer will continuously break down and regrow during Li deposition/stripping processes and is critical to the real growth pattern of reactive metals such as Li. Unfortunately, most models in the literature do not consider the impact of the SEI on the Li dendrite growth mechanisms. Recently, Cheng et al. (Cheng et al. 2015) reviewed the mechanisms of SEI formation and models of SEI structure. The critical factors affecting the SEI formation, such as electrolyte component, temperature, current density, are discussed. An extensive experimental study by Steiger et al. utilizing in situ light microscopy and ex situ SEM analysis concluded, however, that growth of the whisker/needle-like structures (often termed "dendrites") occurs as follows (Steiger et al. 2014b, 2015; Steiger 2015):

- Li needle growth is generally initiated at either the substrate surface or from faceted particles of Li.
- Typically, the needles grow in length, but not in breadth.
- Growth often does not actually occur at the tip, but rather behind an inactive deposit located at the tip (possibly a particle of metal oxide, LiF or other SEI components) (Figs. 2.13, 2.14) (Steiger et al. 2014a).
- Li^+ cation transport through a thin SEI layer results in the overall needle growth which occurs—as just noted—at the needle tip (i.e., deposit-Li interface), at the base (i.e., substrate-Li interface), and at defects resulting in kinks in the needles.
- Extensive shaking/twisting motions transpire during the needle growth process (Steiger et al. 2014a; Brissot et al. 1998; Nishikawa et al. 2007, 2010, 2011, 2012; Nishida et al. 2013; Yamaki et al. 1998)—a characteristic that is readily explained by the differing growth zones (kinks and interfaces) in Steiger's model.

Fig. 2.13 SEM image of Li needle deposits on W (*arrows point* to broadening and/or particles at tips). Reproduced with permission—Copyright 2014, Elsevier (Steiger 2015)

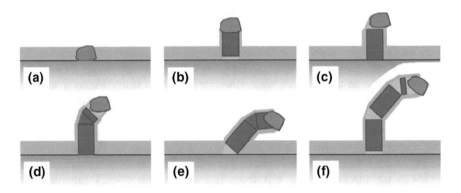

Fig. 2.14 Schematic describing growth of a Li needle: (*green*) SEI—probably mainly organic; (*light blue*) Li insertion areas and (*red*) inert tip. **a** Initial state before Li deposition with an inert inhomogeneity within the SEI, **b** after growing a straight segment by Li insertion at the substrate, **c** after further deposition taking place below the tip, **d** further deposition resulting in a kink, **e** additional Li inserted at the base, causing tilting motions of the whole structure and **f** final structure. All steps proceed by Li insertion into the growing structure. Reproduced with permission —Copyright 2014, Elsevier (Steiger et al. 2014a)

A number of these points were also previously emphasized in publications by Yamaki et al. (1998) and Nishikawa et al. (2011). Note that Steiger's model— which explains the observed Li kinked whisker/needle-like growth patterns quite well—diverges significantly from previous models that have emphasized dendrite formation due to the depletion of Li^+ cations near the electrode surface, field enhancement at the needle ("dendrite") tips, the strong influence of concentration gradients, and stresses that induce needle motion. Figure 2.15 shows SEM images of Li deposits on a stainless steel (SS) electrode after different times during which the electrode was polarized to −150 mV (versus Li/Li^+) in 1 M $LiPF_6$-EC/DMC electrolyte and 1 M N_{1114}TFSI-LiTFSI (IL-based) electrolytes (Stark et al. 2013).

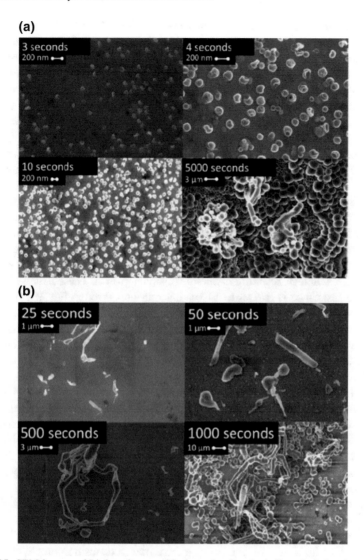

Fig. 2.15 SEM images of Li deposits on a SS electrode after the indicated plating times during which the electrode was polarized to −150 mV (versus Li/Li⁺) in **a** 1 M LiPF₆-EC/DMC electrolyte and **b** 1 M N₁₁₁₄TFSI-LiTFSI (IL-based) electrolytes. Reproduced with permission— Copyright 2013, The Electrochemical Society (Stark et al. 2013)

When the Li deposition rate is slow, the Li nuclei formed initially will merge together as shown in Fig. 2.15a. When the needles grow fast relative to the nucleation points, then tangled fibrous aggregates of the kinked needles tend to result (Fig. 2.15b).

Very different Li deposition is sometimes observed that is referred to as non-dendritic—i.e., the Li has a particulate/nodular structure, which may be fused into aggregated lumps or instead simply clustered together. This then requires a growth model that diverges from the linear needle model noted above. The Li deposition morphology is governed by Li^+ cation mobility, Li^0 (adatom) transport on the Li surface, and perhaps to some extent Li^0 self-diffusion within the bulk of the Li. Most of the focus in many mathematical models developed is on the former (i.e., Li^+ cation mobility within the electrolyte). Jäckle and Groß found that Li^0 atom self-diffusion has relatively high barriers on the most energetically favorable surfaces of the Li body-centered cubic (bcc) crystal and also has a lower tendency (than Mg, for example) to adopt high-coordination configurations (Jäckle and Groß 2014). This reduces the driving force for surface reconstruction from needle-like to nodular shapes. Higher temperatures—with the corresponding increase in Li^0 adatom mobility—would therefore be expected to favor more nodular Li structures (Aryanfar et al. 2015) and this will be shown below to indeed be the case. Another important consideration, however, is the Li^+ cation transport rate through the SEI to the Li growth surface. As noted above, the kinked needle growth may be dictated by favorable insertion of Li^0 adatoms at the tip (often below an inactive particle), the base and at defects resulting in one-dimensional growth. These may be locations where the resistance is lowest to Li^+ cation transport through the SEI or alternatively locations which have the lowest interfacial energy and thus serve as a sink for the adatoms. Transport, however, will be more favorable throughout the entire SEI layer at higher temperature and/or for a more conductive/thinner SEI. Thus, temperature and electrolyte composition strongly impact the Li deposition morphology (Nishikawa et al. 2011).

It will be shown below that such growth actually begins as needles which then thicken into nodules—that is, the one-dimensional, linear elongation which creates the needles transitions at some point to three-dimensional growth at the needle tips and defects or alternatively thickening of the entire needle segment(s) (Steiger et al. 2014b; Steiger 2015; Arakawa et al. 1993; Nishikawa et al. 2011, 2012). A highly aligned Li growth pattern was observed when a robust and uniform SEI layer formed on the surface of the substrate as reported recently by Qian et al. (2015a), i.e., the one-dimensional, linear elongation that creates the needles transitions at some point to three-dimensional growth.

The trace amount of HF derived from the decomposition reaction of $LiPF_6$ with H_2O is electrochemically reduced during the initial Li deposition process to form a uniform and dense LiF-rich SEI layer on the surface of the substrate. This SEI layer is robust and leads to a uniform distribution of the electric field on the substrate surface. In case of low rate deposition, the merged Li particles will favor growth into linear arrays of closely packed nanorods, since neighboring nanorods constrain the formation of the kinked defects, thereby enabling uniform and dendrite-free Li deposition. The surface and cross-section images of the as-deposited, dendrite-free Li films exhibit a self-aligned and highly compact Li nanorod structure as shown in Fig. 2.16, which is consistent with a vivid blue color due to structural coloration. Similar surface morphology was also observed before by several other groups (Stark et al. 2013; Zhang et al. 2014; Qian et al. 2015a).

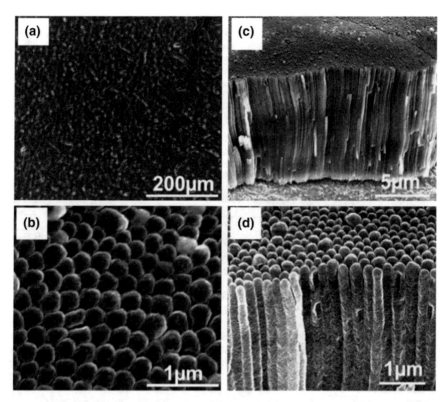

Fig. 2.16 SEM images of the morphologies of Li plated in a 1 M LiPF$_6$-PC electrolyte with 50 ppm H$_2$O additive: **a, b** surface images and **c, d** cross-sectional images. Reproduced with permission—Copyright 2015, Elsevier (Qian et al. 2015a)

2.2.1   "Dead" Lithium

In an early publication in 1974, the effect of aging on the cycling CE of electrodeposited Li was examined using a 1 M LiClO$_4$-PC electrolyte (Selim and Bro 1974). It was noted that the Li was mossy in appearance and that the CE (i.e., stripping/deposition charge ratio) decreased much more rapidly with increasing age of the deposit than did the Li metal content of the deposit (as determined chemically by reacting the Li with water and monitoring the amount of evolved gas generated). For example, after 40 h the CE approached zero, while the chemical analysis indicated that 80 % of the deposited Li remained (but could not be stripped from the electrode). This electrochemically unrecoverable Li has come to be known as "dead" Li (Yoshimatsu et al. 1988; Arakawa et al. 1993; Steiger et al. 2014b). Steiger et al. proposed a mechanism for this that illustrates how the "dead" Li may remain in physical, but not electrical contact with the electrode surface (Fig. 2.17) (Steiger et al. 2014b). This explains why after stripping the electroactive Li from a

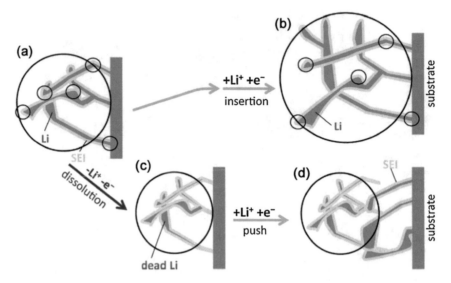

Fig. 2.17 Schematic of proposed growth mechanism for mossy Li. The structure is always covered by an SEI layer: **a** as-deposited, **b** the structure of (**a**) after further electrodeposition. Li atoms are inserted into the metal structure. Points have been marked with black circles to illustrate that the distances between these features generally increase with Li deposition time. The large black oval shape indicates the expansion of the total structure. **c** The structure of (**a**) after a dissolution step. The tips of the structure still contain Li metal ("dead" Li), but are electrically isolated from the substrate, although still being held by the former SEI shell. **d** Structure of (**c**) after an additional electrodeposition step. The top is pushed outward by the new mossy Li growing underneath. Reproduced with permission—Copyright 2014, Elsevier (Steiger et al. 2014b)

previous electrodeposition, new needle-like Li deposits often tend to grow on the electrode surface instead of on the previous (residual) material on the electrode surface (Brissot et al. 1998). This continues until the entire surface of the electrode is covered by the needles and their SEI residue. Thus, the interphasial layer on the surface of the Li may actually grow/accumulate predominantly at the working electrode surface rather than at the interphasial layer/electrolyte interface.

It has been reported from in situ visualization studies for both liquid and polymer electrolytes that dendrites are often not evident during the first few cycles at higher current densities (for which the CE is often relatively high), but then dendrites form and grow in subsequent cycles (often resulting in a continuous decline in the CE with continued cycling) (Dampier and Brummer 1977; Brissot et al. 1998, 1999b). This may be explained by the growth of short new needle-like deposits on the electrode surface during the first few cycles until the electrode is covered in its entirety by the SEI residue and "dead" Li. Then, breakaway Li dendritic structures form while Li deposits on the electrode escape in the following cycles through defects in this resistive layer, culminating in the exposure of protruding kinked needles for which Li^0 adatom addition is more facile, thus resulting in rapid growth of such dendrites.

2.2.2   Interphasial Layer and Formation of Mossy Lithium

For Li-ion batteries in which graphite is used as the anode, the SEI is generally a thin layer of reaction products formed from the degradation of the graphite (near the electrolyte interface), Li, solvent molecules, anions and/or other electrolyte components. Comparable SEI layers may form on the Li metal surface as noted in the discussion above on the needle growth mechanism. But rather than passivating the electrode from further reactions with the electrolyte, the dead Li forms an interphasial porous degradation layer (often termed "mossy" Li) that is much more substantive than the relatively thin SEI layer formed on the graphite surface. Upon cycling, there is thus a transition from a flat, smooth morphology to a rugged structure with a surface interphasial layer that continues to grow in thickness upon cycling (Figs. 2.18, 2.19) (Naoi et al. 1996; Orsini et al. 1998, 1999; López et al. 2009, 2012; Chang et al. 2015; Lv et al. 2015). In comparison, more mossy Li is formed at low current density and more dendrite is formed at high-current density (Orsini et al. 1999). The increased resistance results in increased polarization (i.e., higher absolute voltages for plating/stripping) (Fig. 2.20) and the "dead" Li may mask portions of the electrode (i.e., reduce the active surface area available for

Fig. 2.18 Sectioned Li battery after charging (0.45 mA cm^{-2}): **a** first charge and **b** 14th charge. Reproduced with permission —Copyright 1998, Elsevier (Orsini et al. 1998)

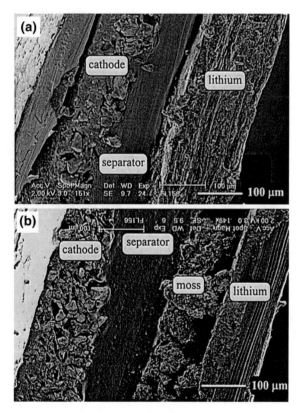

Fig. 2.19 Morphological changes during cycling between two Li electrodes in a 1 M $LiPF_6$-PC electrolyte at a current density of 1 mA cm^{-2}: **a** uncycled, **b** 100 cycles, **c** 200 cycles and **d** 500 cycles. Reproduced with permission—Copyright 2003, Elsevier (Howlett et al. 2003)

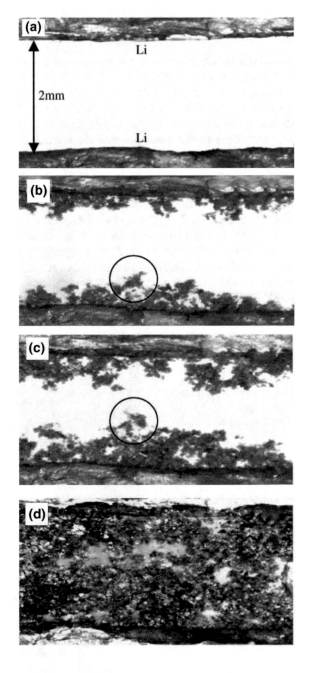

deposition), thus increasing the effective current density which, as will be shown below, often results in more rapid Li consumption and possibly increased dendritic Li growth. This depletion of the electrolyte and electroactive Li due to degradation

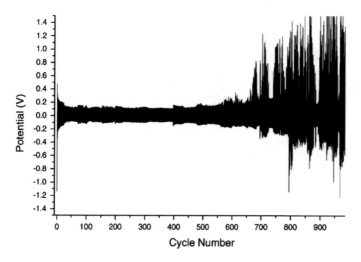

Fig. 2.20 Voltage profile for a Li‖Li cell with a 1 M LiPF$_6$-PC electrolyte cycled at a current density of 1 mA cm^{-2}. Voltage spikes are due to temporary short circuits. Reproduced with permission—Copyright 2003, Elsevier (Howlett et al. 2003)

reactions, as well as the increased interfacial resistance from the porous interphasial layer, has been delineated as a principal cause of cell performance degradation and failure (Fig. 2.21) (Lv et al. 2015), although dendritic short circuiting may occur in some instances (López et al. 2009, 2012; Orsini et al. 1998, 1999; Yoshimatsu et al. 1988). Thus, the large voltage spikes in Fig. 2.20 originate from the highly resistive interphase rather than from short circuiting (which would result in a very low potential between the electrodes). Importantly, for full batteries, the loss of electrolyte (and the corresponding significant impedance increase) may ultimately be the key factor for the deterioration of the cell capacity rather than formation of the interphasial layer (Aurbach et al. 2000).

2.3   Modeling of Lithium Dendrite Growth

Significant work has been done to simulate and predict the growth pattern of Li dendrite growth in the last half-century. When an electrochemical cell containing an electrolyte with a Li salt is sufficiently polarized, Li$^+$ cations near the negative electrode are reduced to Li metal and—depending upon the applied current density and the electrolyte's transport properties—the Li$^+$ cation concentration decreases resulting in anion migration toward the positive electrode until equilibrium is reached. The newly formed Li metal may deposit as a relatively compact layer or in a variety of other morphologies which are often described as dendritic. A recent review of models for dendrite initiation/propagation described three classifications (Li et al. 2014b):

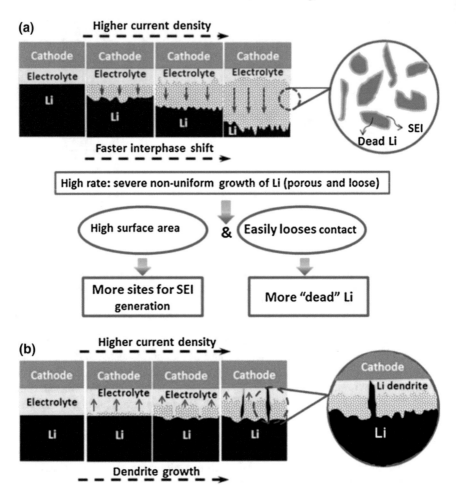

Fig. 2.21 Failure mechanism of the Li anode at high charge current densities: **a** schematic illustration of the failure mechanism and **b** conventional understanding of the dendrite-related failure mechanism for Li batteries. Reproduced with permission—Copyright 2014, Wiley-VCH (Lv et al. 2015)

1. Surface Tension Model—This model finds that electrodeposition is more rapid on projections rather than planar surfaces because spherical rather than linear diffusion dominates the mass transport of the active species. A larger spherical diffusion flux results in a narrowing of the growing dendrite tip; thus surface forces and mass transport dominate the kinetics governing dendrite growth (Monroe and Newman 2003, 2004, 2005; Barton and Bockris 1962; Diggle et al. 1969).

2. Brownian Statistical Simulation Model—This model focuses on the competition between ion transport and reductive deposition. When deposition probability is

low, Li^+ cation transport dominates and the ions penetrate close to the substrate resulting in a dense plating morphology with a low tendency for dendrite formation. This is contrasted with the case of high-deposition probability (relative to the cation transport), which increases the rate of deposition at the tip of projecting growth structures thus increasing the tendency for dendrite formation (Mayers et al. 2012; Voss and Tomkiewicz 1985; Magan et al. 2003).

3. Chazalviel Electromigration-Limited Model—This model for dendrite growth is often cited: at a time τ called the Sand time, the Li^+ cation concentration drops to zero at the negative electrode with high currents, causing the potential to diverge; this results in an instability at the interface from inhomogeneities in the distribution of the surface potential, which creates a localized electric field that leads to dendrite growth. The dendrite initiation time thus corresponds to the buildup of the space charge and the dendrite propagation velocity is tied to the transport of the anions (Brissot et al. 1998; Chazalviel 1990).

In addition to these, several new models have recently been proposed to describe Li plating/growth (Akolkar 2013, 2014; Chen et al. 2015; Cogswell 2015; Tang et al. 2009; Aryanfar et al. 2014). More details on a few of these important models of dendrite growth will be discussed in this section.

2.3.1 General Models

In the field of electrodeposition, metal "dendrites" are a common phenomenon. At a given electrodeposition condition, many metals, such as zinc (Zn), copper (Cu), silver (Ag), and tin (Sn), were reported to exhibit ramified morphologies (Chazalviel 1990). Fractal deposits including needle-like, snowflake-like, tree-like, bush-like, moss-like, and whisker-like structures are all referred to as dendrites in this review. Extensive work has been done on the dendrite formation mechanism during electrodeposition of Zn and Cu. Various strategies have been unitized to suppress dendrite growth in these processes (Sawada et al. 1986; Diggle et al. 1969). The reported methods to suppress Zn dendrites include special separators, alternating current or pulsed charging, and additives in the electrolytes. The last methods can be further divided into three categories: electrode structural modifiers, metallic additives, and organic additives (Lan et al. 2007). Several factors such as Zn concentration, complexing agents, anions, and additives may modify the texture and morphology of Zn electrodeposited coatings (Mendoza-Huízar et al. 2009). Recently, Aaboubi et al. (2011) investigated the effect of tartaric acid on Zn electrodeposition from a sulfate plating bath by electrochemical impedance spectroscopy (EIS), stationary polarization curves, XRD, and SEM imagery. The study shows that it is possible to obtain homogeneous, compact, and dendrite-free Zn deposits from sulfate solutions containing tartaric acid. Miyazaki et al. (2012)

also reported suppression of dendrite formation in metallic Zn deposition using zinc oxide electrodes modified with an anion-exchange ionomer (AEI). These improvements are explained by selective ion permeation through the AEI films. These approaches on Zn dendrite suppression and the techniques used in the investigation of Zn dendrite suppression and the technics used in the investigation of Zn dendrite growth are very useful for the investigation and prevention of Li dendrite growth.

Although most electrodepositions are a one-time-only process, in a rechargeable Li metal battery, Li metal needs to be plated on or stripped from substrates repeatedly during charge/discharge processes. As a result, Li dendrites will accumulate on the anode and finally lead to many serious problems (see Fig. 1.1) that hamper the practical application of rechargeable Li metal batteries. Therefore, a good understanding of the mechanism of Li dendrite formation and growth is critical to mitigate or further eliminate Li dendrites.

Many groups have simulated the Li dendrite formation and growth process, and proposed several meaningful and fundamental models in the last forty years. In order to simplify the simulation conditions, most models were based on a binary electrolyte with a Li salt and polymer; for example, $LiClO_4$ or $LiN(SO_2CF_3)_2$ (LiTFSI) in polyethylene oxide (PEO). In the open circuit condition, the electrolyte is in a steady state without an ionic concentration gradient; under polarization, the Li^+ and anion will diverge and transfer to the negative and positive electrode, respectively. Li^+ will obtain an electron and plate on the negative electrode. The speed of Li deposition or consumption of Li^+ depends on the applied current density. Although the depletion of Li^+ can be macroscopically compensated by the supply of Li^+ from the positive electrode, the microscopic ionic distribution near the negative electrode dramatically affects the deposit's morphology. Therefore, a basic model to simulate Li dendrite starts from the calculation of the concentration gradient in the Li symmetric cell under polarization. Brissot and Chazalviel et al. described the concentration gradient in a cell with a small inter-electrode distance using the following equation (Rosso et al. 2006; Brissot et al. 1999b):

$$\frac{\partial C}{\partial x}(x) = \frac{J\mu_a}{eD(\mu_a + \mu_{Li^+})} \tag{2.1}$$

where J is the effective electrode current density, D is the ambipolar diffusion coefficient, e is the electronic charge, and μ_a and μ_{Li}^+ are the anionic and Li^+ mobilities. From Eq. (2.1), two different conditions for a symmetrical Li/PEO/Li cell can be anticipated, with the inter-electrode distance L and initial Li salt concentration C_o:

(a) If $dC/dx < 2C_o/L$, the ionic concentration evolves to a steady state where the concentration gradient is constant and varies almost linearly from $C_a = C_o - \Delta C_a$ at the negative electrode to $C_{Li^+} = C_o + \Delta C_{Li^+}$ at the positive electrode, where

$$-\Delta C_a \approx -\Delta C_{Li^+} \approx \frac{\mu_a}{\mu_a + \mu_{Li^+}} \frac{JL}{eD} \tag{2.2}$$

(b) If $dC/dx > 2C_0/L$, the ionic concentration goes to zero at the negative electrode at a time called "Sand's time" τ, which varies as

$$\tau = \pi D \left(\frac{eC_0}{2Jt_a}\right)^2 \tag{2.3}$$

$$t_a \approx 1 - t_{Li^+} = \frac{\mu_a}{\mu_a + \mu_{Li^+}} \tag{2.4}$$

where t_a and t_{Li}^+ represent the anionic and Li^+ transference number, respectively. Chazalviel indicated that the anionic and Li^+ concentrations exhibit different behaviors at the Sand time, leading to an excess of positive charge at the negative electrode. This behavior will result in a local space charge, form a large electric field, and lead to nucleation and growth of Li dendrite. The results of their simulations and experiments confirmed the concentration gradient and the occurrence of dendrites very close to Sand's time (Brissot et al. 1998, 1999a). Chazalviel (1990) also predicted that the dendrite will grow at the velocity of

$$v = -\mu_a E \tag{2.5}$$

where E is the electric field strength.

Monroe and Newman developed the general model describing dendrite growth under galvanostatic conditions applicable to liquid electrolytes (Monroe and Newman 2003). They adopted the Barton and Bockris method (Barton and Bockris 1962) with the addition of thermodynamic reference points and the fact that concentration and potential are not constant during the course of dendrite growth. They calculated the concentration and potential distribution in the cell at different time intervals. It was demonstrated that the dendrite growth rate increases across the electrolyte and depends greatly on the applied current density; this will be discussed in more detail in the following section.

Rosso et al. reported a systematic study on the evolution of Li dendrites in LiTFSI-PEO electrolyte involving theoretical calculations as well as experimental data (Rosso et al. 2006). They demonstrated that even though the formation of dendrites has little effect on overall impedance, it significantly decreases interfacial impedance. Based on the impedance data, the value of resistance due to the dendrites could be calculated. In addition, it was observed that dendrites can burn out like a fuse; that is, when the first dendrite reaches the opposite electrode it shorts the circuit and the current density passing through this single dendrite becomes high enough to melt it. Thus, the final short circuit occurs only when the major front of dendrites eventually reaches the opposite electrode.

Although the formation of contiguous and conducting Li dendrites in batteries is often called "dendritic growth," there are actually several modes of formation and growth: dendrites, whiskers, and "others." The true dendrite grows from a Li metal surface in a nonaqueous electrolyte by adding material to its tip. The nutrient source is the Li in the electrolyte. Classical models of dendrite growth gave solutions in the form of the tip radius times its velocity, which has the units of diffusivity. Recent electrochemical continuum models (Rosso et al. 2001) and experiments (Bhattacharyya et al. 2010) for Li batteries have found that the dendrite growth is controlled by the tip surface energy, always accelerates across the cell under all conditions, and can be partially mitigated by lowering the limiting current or increasing the cell thickness. The latter two conditions limit the performance of the battery. A second type of growth has been simulated in some Li battery experiments. If the nutrient supply is drawn from the Li metal sheet, growth occurs at the base of a "whisker." Yamaki (1998) analyzed the stress-assisted whisker growth through cracks in a protective layer (i.e., the separator) on the surface of the Li anode.

2.3.2 Effect of Current Density

It is well known that the effective current density during Li deposition/stripping has a significant impact on the dendrite formation and growth. Generally, low current density results in relatively stable cycling, and conversely, high current density accelerates the degradation process of rechargeable Li metal batteries. The equation of Sand's time indicates that the dendrite initiation time is proportional to J^{-2}, which indicates that high current density greatly accelerates the formation of Li dendrites. It is worth noting that there are some results showing $\tau \sim J^{-1.25}$ rather than $\tau \sim J^{-2}$ dependence, as reported by (Liu et al. 2010). They attributed this deviation to the local fluctuations of current density. Moreover, the ionic liquid (IL) used in their work acted as a supporting electrolyte. In fact, this is a ternary electrolyte rather than a binary one as assumed by Chazalviel's model. In the model developed by Monroe and Newman (2003) using liquid electrolyte, tip growth rate (v_{tip}) can be expressed as

$$v_{tip} = \frac{J_n V}{F} \tag{2.6}$$

where J_n is the effective current density normal to the dendrite (hemispherical) tip, V is the molar volume of Li and F is Faraday's constant. This equation suggests that the dendrite growth rate is proportional to J_n. Based on Eq. (2.3) and (2.6), the dendrite initiation time could be delayed and the dendrite growth rate could be slowed down if the effective current density could be decreased. By applying a smaller current density, a smoother surface and improved cycling life have been obtained (Kanamura et al. 1996; Aurbach et al. 2000; Zinigrad et al. 2001; Wang et al. 2000; Crowther and West 2008; Gireaud et al. 2006).

According to Chazalviel's model, an applied current density leads to an ion concentration gradient—high-current density results in near-zero ion concentration at the negative electrode and the formation of Li dendrites at Sand's time, low current density leads to a minimal and stable ion concentration gradient so no Li dendrites form in this condition. The crossover between the two regimes is the limiting current density

$$J^* = 2eC_oD/t_aL \tag{2.7}$$

where L is the inter-electrode distance, t_a is the anionic transport number. When the current density is low or the inter-electrode distance L *is small*, there is in principle no Sand behavior and the concentration variation should be small. However, experimental results clearly indicate there are still Li dendrites, just not as serious as those at high-current density. Rosso et al. (2001) and Teyssot et al. (2005) attributed the formation of dendrites to local nonuniformity of the Li/electrolyte interface, which leads to a large concentration variance even in the depleted zone close to the conditions of Chazalviel's model. Brissot et al. (1998) confirmed this experimentally in a Li| LiTFSI-PEO|Li cell although individual dendrites could deviate from the predicted growth rate. It was demonstrated that at high current densities (when Li deposition becomes diffusion controlled), the onset of dendrite formation matched Sand's time (zero ion concentration). However, dendrites started to grow earlier with cycling, apparently because of the created defects. At low current densities (i.e., when concentration variations were low), dendrites were also observed in the form of elongated metal filaments (higher aspect ratio), which could be a result of local inhomogeneity. Growth velocities followed Chazalviel's model (Chazalviel 1990) in both cases. It was later shown by Rosso et al. (2001) that the time of dendrite appearance at low current densities is also proportional to the power of current density even though Sand's behavior was not expected. It was proposed that the specific properties of LiTFSI-PEO electrolyte cause destabilization of the concentration distribution along the electrode surface. A direct relation between dendritic growth and concentration gradient was clearly demonstrated by employing three independent techniques to measure ion concentrations in the vicinity of dendrites (Brissot et al. 1999a).

In addition to the value of current density, the charging styles, galvanostat, or pulse also significantly affects Li dendrite formation and growth. Recent work by Miller and coauthors reported that pulsed charging can effectively suppress Li dendrite formation by as much as 96 %. They proposed a coarse-grained lattice model to explain the mechanism of pulsed charging and revealed that dendrite formation arose from a competition between the time scales of Li^+ diffusion and reduction at the anode, with lower overpotential and shorter electrode pulse durations shifting this competition in favor of lower dendrite formation probability (Mayers et al. 2012).

2.3.3 Importance of Interfacial Elastic Strength

Monroe and Newman (2005) further employed linear elasticity theory to develop a kinetic model describing how mechanical properties of polymer electrolytes (shear modulus and Poisson's ratio) affect roughness on the Li interface. The interface was subjected to a regime of small-amplitude two-dimensional (2D) perturbations. Analytic solutions with specific boundary conditions allowed computing deformation profiles. Compressive stress, deformation stress, and surface tension at the elastic Li interface were calculated as a next step. Incorporation of these parameters into the model gave a prediction of the distribution of exchange current density along the electrode surface. Finally, it was possible to verify that the mechanical properties of the polymer electrolyte stopped amplification of the dendrite growth. It turned out that dendrite suppression can be achieved when the shear modulus of the electrolyte is about twice that of the Li anode ($\sim 10^9$ Pa); that is, at least three orders of magnitude higher than that of the studied PEO.

As in many other fields, all of the models discussed above have their limitations. For example, the dendrite growth velocity proposed by Monroe and Newman (2003) (Eq. (2.6)) was derived from the growth of a single dendrite without considering the interaction between neighboring dendrites. It was also stated that the Chazalviel theory (Chazalviel 1990) has limited application in real batteries because it applies at currents higher than the limiting current. However, Rosso et al. (2001) and Teyssot et al. (2005) concluded that the Chazalviel model can be extended to low currents due to the nanoscale inhomogeneity in concentration, at least in the case of PEO-based electrolytes. Even though these models still include many simplifications and limitations, they have established a solid foundation for the nucleation and growth mechanisms of dendrites. More importantly, as we will discuss later in this book, several general predications of these models have been used successfully to identify new approaches to suppress dendrite growth, especially during Li metal deposition; for example, using an anode with large surface area to reduce the effective current density, developing a single ion conductor to enhance Li^+ transference number, developing an electrolyte with strong shear modulus, and adding supporting electrolyte. These approaches will be discussed in detail in Chap. 3.

References

Aaboubi O, Douglade J, Abenaqui X, Boumedmed R, VonHoff J (2011) Influence of tartaric acid on zinc electrodeposition from sulphate bath. Electrochim Acta 56(23):7885–7889

Akolkar R (2013) Mathematical model of the dendritic growth during lithium electrodeposition. J Power Sources 232:23–28. doi:10.1016/j.jpowsour.2013.01.014

Akolkar R (2014) Modeling dendrite growth during lithium electrodeposition at sub-ambient temperature. J Power Sources 246:84–89. doi:10.1016/j.jpowsour.2013.07.056

Arai J, Okada Y, Sugiyama T, Izuka M, Gotoh K, Takeda K (2015) In situ solid state 7Li NMR observations of lithium metal deposition during overcharge in lithium ion batteries. J Electrochem Soc 162(6):A952–A958. doi:10.1149/2.0411506jes

Arakawa M, Tobishima S-I, Nemoto Y, Ichimura M (1993) Lithium electrode cycleability and morphology dependence on current density. J Power Sources 43–44:27–35

Aryanfar A, Brooks D, Merinov BV, Goddard WA III, Colussi AJ, Hoffmann MR (2014) Dynamics of lithium dendrite growth and inhibition: pulse charging experiments and monte carlo calculations. J Phys Chem Lett 5(10):1721–1726. doi:10.1021/jz500207a

Aryanfar A, Brooks DJ, Colussi AJ, Merinov BV, Goddard WA III, Hoffmann MR (2015) Thermal relaxation of lithium dendrites. Phys Chem Chem Phys 17(12):8000–8005. doi:10.1039/c4cp05786d

Aurbach D, Cohen Y (1996) The application of atomic force microscopy for the study of Li deposition processes. J Electrochem Soc 143(11):3525–3532

Aurbach D, Cohen Y (1997) Morphological studies of Li deposition processes in LiAsF6/PC solutions by in situ atomic force microscopy. J Electrochem Soc 144(10):3355–3360

Aurbach D, Daroux ML, Faguy PW, Yeager E (1987) Identification of surface films formed on lithium in propylene carbonate solutions. J Electrochem Soc 134(7):1611–1620

Aurbach D, Gofer Y, Langzam J (1989) The correlation between surface chemistry, surface morphology, and cycling efficiency of lithium electrodes in a few polar aprotic systems. J Electrochem Soc 136(11):3198–3205

Aurbach D, Youngman O, Dan P (1990a) The electrochemical behavior of 1,3-Dioxolane-LiClO4 solutions—II. Contaminated solutions. Electrochim Acta 35(3):639–655

Aurbach D, Youngman O, Gofer Y, Meitav A (1990b) The electrochemical behavior of 1,3-Dioxolane-LiClO4 solutions—I. Uncontaminated Solutions. Electrochim Acta 35(3):625–638

Aurbach D, Daroux M, McDougall G, Yeager EB (1993) Spectroscopic studies of lithium in an ultrahigh vacuum system. J Electroanal Chem 358:63–76. doi:10.1016/0022-0728(93)80431-G

Aurbach D, Weissman I, Zaban A, Chusid O (1994) Correlation between surface chemistry, morphology, cycling efficiency and interfacial properties of Li electrodes in solutions containing different Li salts. Electrochim Acta 39:51–71

Aurbach D, Zaban A, Gofer Y, Ely YE, Weissman I, Chusid O, Abramson O (1995) Recent studies of the lithium-liquid electrolyte interface. electrochemical, morphological and spectral studies of a few important systems. J Power Sources 54:76–84

Aurbach D, Zaban A, Ein-Eli Y, Weissman I, Chusid O, Markovsky B, Levi M, Levi E, Schechter A, Granot E (1997) Recent studies on the correlation between surface chemistry, morphology, three-dimensional structures and performance of Li and Li-C intercalation anodes in several important electrolyte systems. J Power Sources 68:91–98

Aurbach D, Weissman I, Yamin H, Elster E (1998) The correlation between charge/discharge rates and morphology, surface chemistry, and performance of Li electrodes and the connection to cycle life of practical batteries. J Electrochem Soc 145(5):1421–1426

Aurbach D, Markovsky B, Levi MD, Levi E, Schechter A, Moshkovich M, Cohen Y (1999) New insights into the interactions between electrode materials and electrolyte solutions for advanced nonaqueous batteries. J Power Sources 81–82:95–111

Aurbach D, Zinigrad E, Teller H, Dan P (2000) Factors which limit the cycle life of rechargeable lithium (metal) batteries. J Electrochem Soc 147:1274–1279

Barton JL, Bockris JOM (1962) The electrolytic growth of dendrites from ionic solutions. Proc R Soc Lond A 268:485–505

Besenhard JO, Gürtler J, Komenda P, Paxinos A (1987) Corrosion protection of secondary lithium electrodes in organic electrolytes. J Power Sources 20:253–258

Bhattacharyya R, Key B, Chen H, Best AS, Hollenkamp AF, Grey CP (2010) In situ NMR observation of the formation of metallic lithium microstructures in lithium batteries. Nat Mater 9(6):504–510. doi:10.1038/nmat2764

Bieker G, Winter M, Bieker P (2015) Electrochemical in situ investigations of SEI and dendrite formation on the lithium metal anode. Phys Chem Chem Phys 17(14):8670–8679. doi:10.1039/c4cp05865h

Brissot C, Rosso M, Chazalviel JN, Baudry P, Lascaud S (1998) In situ study of dendritic growth in lithium/PEO-salt/lithium cells. Electrochim Acta 43(10–11):1569–1574

Brissot C, Rosso M, Chazalviel JN, Lascaud S (1999a) In situ concentration cartography in the neighborhood of dendrites growing in lithium/polymer-electrolyte/lithium cells. J Electrochem Soc 146(12):4393–4400

Brissot C, Rosso M, Chazalviel JN, Lascaud S (1999b) Dendritic growth mechanisms in lithium/polymer cells. J Power Sources 81–82:925–929. doi:10.1016/S0378-7753(98)00242-0

Chandrashekar S, Trease NM, Chang HJ, Du L-S, Grey CP, Jerschow A (2012) 7Li MRI of Li batteries reveals location of microstructural lithium. Nat Mater 11(4):311–315. doi:http://www.nature.com/nmat/journal/v11/n4/abs/nmat3246.html#supplementary-information

Chang HJ, Trease NM, Ilott AJ, Zeng D, Du L-S, Jerschow A, Grey CP (2015) Investigating Li microstructure formation on Li anodes for lithium batteries by in situ 6Li/7Li NMR and SEM. J Phys Chem C 119(29):16443–16451. doi:10.1021/acs.jpcc.5b03396

Chazalviel JN (1990) Electrochemical aspects of the generation of ramified metallic electrodeposits. Phys Rev A 42(12):7355–7367. doi:10.1103/PhysRevA.42.7355

Chen L, Zhang HW, Liang LY, Liu Z, Qi Y, Lu P, Chen J, Chen L-Q (2015) Modulation of dendritic patterns during electrodeposition: a nonlinear phase-field model. J Power Sources 300:376–385. doi:10.1016/j.jpowsour.2015.09.055

Cheng X-B, Zhang R, Zhao C-Z, Wei F, Zhang J-G, Zhang Q (2015) A review of solid electrolyte interphases on lithium metal anode. Adv Sci: 1500213

Choi N-S, Koo B, Yeon J-T, Lee KT, Kim D-W (2011) Effect of a novel amphipathic ionic liquid on lithium deposition in gel polymer electrolytes. Electrochim Acta 56(21):7249–7255. doi:10.1016/j.electacta.2011.06.058

Cogswell DA (2015) Quantitative phase-field modeling of dendritic electrodeposition. Phys Rev E 92(1):011301. doi:10.1103/PhysRevE.92.011301

Cohen YS, Cohen Y, Aurbach D (2000) Micromorphological studies of lithium electrodes in alkyl carbonate solutions using in situ atomic force microscopy. J Phys Chem B 104:12282–12291

Crowther O, West AC (2008) Effect of electrolyte composition on lithium dendrite growth. J Electrochem Soc 155(11):A806–A811. doi:10.1149/1.2969424

Dampier FW, Brummer SB (1977) The cycling behavior of the lithium electrode in LiAsF6/methyl acetate solutions. Electrochim Acta 22:1339–1345

Deng X, Hu MY, Wei X, Wang W, Chen Z, Liu J, Hu JZ (2015) Natural abundance [17]O nuclear magnetic resonance and computational modeling studies of lithium based liquid. J Power Sources 285:146–155. doi:10.1016/j.jpowsour.2015.03.091

Dey AN (1976) S.E.M. studies of the Li-film growth and the voltage-delay phenomenon associated with the lithium-thionyl chloride inorganic electrolyte system. Electrochim Acta 21 (5):377–382

Diggle JW, Despic AR, Bockris JOM (1969) The mechanism of the dendritic electrocrystallization of zinc. J Electrochem Soc 112:1503–1514

Ding F, Xu W, Graff GL, Zhang J, Sushko ML, Chen X, Shao Y, Engelhard MH, Nie Z, Xiao J, Liu X, Sushko PV, Liu J, Zhang J-G (2013) Dendrite-free lithium deposition via self-healing electrostatic shield mechanism. J Am Chem Soc 135(11):4450–4456. doi:10.1021/ja312241y

Ding F, Xu W, Chen X, Zhang J, Shao Y, Engelhard MH, Zhang Y, Blake TA, Graff GL, Liu X, Zhang J-G (2014) Effects of cesium cations in lithium deposition via self-healing electrostatic shield mechanism. J Phys Chem C 118(8):4043–4049. doi:10.1021/jp4127754

Dollé M, Sannier L, Beaudoin B, Trentin M, Tarascon J-M (2002) Live scanning electron microscope observations of dendritic growth in lithium/polymer cells. Electrochem Solid-State Lett 5(12):A286–A289. doi:10.1149/1.1519970

Fujieda T, Yamamoto N, Saito K, Ishibashi T, Honjo M, Koike S, Wakabayashi N, Higuchi S (1994) Surface of lithium electrodes prepared in Ar + CO_2 Gas. J Power Sources 52:197–200. doi:10.1016/0378-7753(94)01961-4

Ghassemi H, Au M, Chen N, Heiden PA, Yassar RS (2011) Real-time observation of lithium fibers growth inside a nanoscale lithium-ion battery. Appl Phys Lett 99(12):123113. doi:10.1063/1.3643035

Gireaud L, Grugeon S, Laruelle S, Yrieix B, Tarascon JM (2006) Lithium metal stripping/plating mechanisms studies: a metallurgical approach. Electrochem Commun 8(10):1639–1649. doi:10.1016/j.elecom.2006.07.037

Gu M, Parent LR, Mehdi BL, Unocic RR, McDowell MT, Sacci RL, Xu W, Connell JG, Xu P, Abellan P, Chen X, Zhang Y, Perea DE, Evans JE, Lauhon LJ, Zhang JG, Liu J, Browning ND, Cui Y, Arslan I, Wang CM (2013) Demonstration of an electrochemical liquid cell for operando transmission electron microscopy observation of the lithiation/delithiation behavior of Si nanowire battery anodes. Nano Lett 13(12):6106–6112. doi:10.1021/nl403402q

Harry KJ, Hallinan DT, Parkinson DY, MacDowell AA, Balsara NP (2014) Detection of subsurface structures underneath dendrites formed on cycled lithium metal electrodes. Nat Mater 13(1):69–73. doi:10.1038/nmat3793

Harry KJ, Liao X, Parkinson DY, Minor aM, Balsaraa NP (2015) Electrochemical Deposition and Stripping Behavior of Lithium Metal across a Rigid Block Copolymer Electrolyte Membrane. J Electrochem Soc 162(14):A2699–A2706

Hernandez-Maya R, Rosas O, Saunders J, Castaneda H (2015) Dynamic characterization of dendrite deposition and growth in Li-surface by electrochemical impedance spectroscopy. J Electrochem Soc 162(4):A687–A696. doi:10.1149/2.0561504jes

Howlett PC, MacFarlane DR, Hollenkamp AF (2003) A sealed optical cell for the study of lithium-electrode|electrolyte interfaces. J Power Sources 114(2):277–284. doi:10.1016/s0378-7753(02)00603-1

Huang JY, Zhong L, Wang CM, Sullivan JP, Xu W, Zhang LQ, Mao SX, Hudak NS, Liu XH, Subramanian A, Fan H, Qi L, Kushima A, Li J (2010) In situ observation of the electrochemical lithiation of a single SnO_2 nanowire electrode. Science 330:1515–1520

Ishikawa M, Morita M, Matsuda Y (1997) In situ scanning vibrating electrode technique for lithium metal anodes. J Power Sources 68:501–505

Jäckle M, Groß A (2014) Microscopic properties of lithium, sodium, and magnesium battery anode materials related to possible dendrite growth. J Chem Phys 141(17):174710. doi:10.1063/1.4901055

Kanamura K, Tamura H, Shiraishi S, Takehara Z-I (1995a) Morphology and chemical compositions of surface films of lithium deposited on a Ni substrate in nonaqueous electrolytes. J Electroanal Chem 394:49–62

Kanamura K, Tamura H, Shiraishi S, Takehara Z-I (1995b) XPS analysis of lithium surfaces following immersion in various solvents containing LiBF4. J Electrochem Soc 142(2):340–347

Kanamura K, Shiraishi S, Takeharo Z-I (1996) Electrochemical deposition of very smooth lithium using nonaqueous electrolytes containing HF. J Electrochem Soc 143(7):2187–2197

Kanamura K, Takezawa H, Shiraishi S, Takehara Z-I (1997) Chemical reaction of lithium surface during immersion in $LiClO_4$ or $LiPF_6$/DEC electrolyte. J Electrochem Soc 144(6):1900–1906

Kominato A, Yasukawa E, Sato N, Ijuuin T, Asahina H, Mori S (1997) Analysis of surface films on lithium in various organic electrolytes. J Power Sources 68:471–475

Lan CJ, Lee CY, Chin TS (2007) Tetra-alkyl ammonium hydroxides as inhibitors of Zn dendrite in Zn-based secondary batteries. Electrochim Acta 52(17):5407–54116

Lee Y-G, Kyhm K, Choi N-S, Ryu KS (2006) Submicroporous/microporous and compatible/incompatible multi-functional dual-layer polymer electrolytes and their interfacial characteristics with lithium metal anode. J Power Sources 163(1):264–268. doi:10.1016/j.jpowsour.2006.05.008

Lee YM, Seo JE, Lee Y-G, Lee SH, Cho KY, Park J-K (2007) Effects of triacetoxyvinylsilane as SEI layer additive on electrochemical performance of lithium metal secondary battery. Electrochem Solid-State Lett 10(9):A216–A219. doi:10.1149/1.2750439

Li W, Zheng H, Chu G, Luo F, Zheng J, Xiao D, Li X, Gu L, Li H, Wei X, Chen Q, Chen L (2014a) Effect of electrochemical dissolution and deposition order on lithium dendrite formation: a top view investigation. Faraday Discuss 176:109–124. doi:10.1039/c4fd00124a

Li Z, Huang J, Liaw BY, Metzler V, Zhang J (2014b) A review of lithium deposition in lithium-ion and lithium metal secondary batteries. J Power Sources 254:168–182. doi:10.1016/j.jpowsour.2013.12.099

Liu S, Imanishi N, Zhang T, Hirano A, Takeda Y, Yamamoto O, Yang J (2010) Lithium dendrite formation in Li/Poly(ethylene oxide)–lithium Bis(trifluoromethanesulfonyl)imide and N-Methyl-N-propylpiperidinium Bis(trifluoromethanesulfonyl)imide/Li cells. J Electrochem Soc 157(10):A1092–A1098. doi:10.1149/1.3473790

Liu XH, Zhong L, Zhang LQ, Kushima A, Mao SX, Li J, Ye ZZ, Sullivan JP, Huang JY (2011) Lithium fiber growth on the anode in a nanowire lithium ion battery during charging. Appl Phys Lett 98(18):183107. doi:10.1063/1.3585655

López CM, Vaughey JT, Dees DW (2009) Morphological transitions on lithium metal anodes. J Electrochem Soc 156(9):A726–A729. doi:10.1149/1.3158548

López CM, Vaughey JT, Dees DW (2012) Insights into the role of interphasial morphology on the electrochemical performance of lithium electrodes. J Electrochem Soc 159(6):A873–A886. doi:10.1149/2.100206jes

Lv D, Shao Y, Lozano T, Bennett WD, Graff GL, Polzin B, Zhang J-G, Engelhard MH, Saenz NT, Henderson WA, Bhattacharya P, Liu J, Xiao J (2015) Failure mechanism for fast-charged lithium metal batteries with liquid electrolytes. Adv Energy Mater 5(3):1400993. doi:10.1002/aenm.201400993

Magan RV, Sureshkumar R, Lin B (2003) Influence of surface reaction rate on the size dispersion of interfacial nanostructures. J Phys Chem B 107:10513–10520

Matsuda Y, Ishikawa M, Yoshitake S, Morita M (1995) Characterization of the lithium-organic electrolyte interface containing inorganic and organic additives by in situ techniques. J Power Sources 54:301–305

Mayers MZ, Kaminski JW, Miller TF (2012) Suppression of dendrite formation via pulse charging in rechargeable lithium metal batteries. J Phys Chem C 116(50):26214–26221. doi:10.1021/jp309321w

Mehdi BL, Gu M, Parent LR, Xu W, Nasybulin EN, Chen X, Unocic RR, Xu P, Welch DA, Abellan P, Zhang JG, Liu J, Wang CM, Arslan I, Evans J, Browning ND (2014) In-situ electrochemical transmission electron microscopy for battery research. Microsc Microanal 20 (2):484–492. doi:10.1017/S1431927614000488

Mehdi BL, Qian J, Nasybulin E, Park C, Welch DA, Faller R, Mehta H, Henderson WA, Xu W, Wang CM, Evans JE, Liu J, Zhang JG, Mueller KT, Browning ND (2015) Observation and quantification of nanoscale processes in lithium batteries by operando electrochemical (S)TEM. Nano Lett 15(3):2168–2173. doi:10.1021/acs.nanolett.5b00175

Mendoza-Huízar LH, Rios-Reyes CH, Gómez-Villegas MG (2009) Zinc electrodeposition from chloride solutions onto glassy carbon electrode. J Mex Chem Soc 53(4):243–247

Miyazaki K, Lee YS, Fukutsuka T, Abe T (2012) Suppression of dendrite formation of zinc electrodes by the modification of anion-exchange ionomer. Electrochemistry 80(10):725–727. doi:10.5796/electrochemistry.80.725

Mogi R, Inaba M, Iriyama Y, Abe T, Ogumi Z (2002a) In situ atomic force microscopy study on lithium deposition on nickel substrates at elevated temperatures. J Electrochem Soc 149(4):A385–A390. doi:10.1149/1.1454138

Mogi R, Inaba M, Iriyama Y, Abe T, Ogumi Z (2002b) Surface film formation on nickel electrodes in a propylene carbonate solution at elevated temperatures. J Power Sources 108:163–173

Mogi R, Inaba M, Jeong S-K, Iriyama Y, Abe T, Ogumi Z (2002c) Effects of some organic additives on lithium deposition in propylene carbonate. J Electrochem Soc 149(12):A1578–A1583. doi:10.1149/1.1516770

Monroe C, Newman J (2003) Dendrite growth in lithium/polymer systems: a propagation model for liquid electrolytes under galvanostatic conditions. J Electrochem Soc 150(10):A1377–A1384. doi:10.1149/1.1606686

Monroe C, Newman J (2004) The effect of interfacial deformation on electrodeposition kinetics. J Electrochem Soc 151(6):A880–A886. doi:10.1149/1.1710893

Monroe C, Newman J (2005) The impact of elastic deformation on deposition kinetics at lithium/polymer interfaces. J Electrochem Soc 152(2):A396–A404. doi:10.1149/1.1850854

Morigaki K-I (2002) Analysis of the interface between lithium and organic electrolyte solution. J Power Sources 104:13–23

Morigaki K-I, Ohta A (1998) Analysis of the surface of lithium in organic electrolyte by atomic force microscopy, fourier transform infrared spectroscopy and scanning auger electron microscopy. J Power Sources 76:159–166

Nagao M, Hayashi A, Tatsumisago M, Kanetsuku T, Tsuda T, Kuwabata S (2013) In situ SEM study of a lithium deposition and dissolution mechanism in a bulk-type solid-state cell with a Li2S-P2S5 solid electrolyte. Phys Chem Chem Phys 15(42):18600–18606. doi:10.1039/c3cp51059j

Naoi K, Mori M, Shinagawa Y (1996) Study of deposition and dissolution processes of lithium in carbonate-based solutions by means of the quartz-crystal microbalance. J Electrochem Soc 143 (8):2517–2522

Naoi K, Mori M, Naruoka Y, Lamanna WM, Atanasoski R (1999) The surface film formed on a lithium metal electrode in a new imide electrolyte, lithium Bis(perfluoroethylsulfonylimide) [LiN(C2F5SO2)2]. J Electrochem Soc 146(2):462–469

Naudin C, Bruneel JL, Chami M, Desbat B, Grondin J, Lassègues JC, Servant L (2003) Characterization of the lithium surface by infrared and raman spectroscopies. J Power Sources 124(2):518–525. doi:10.1016/s0378-7753(03)00798-5

Nazri G, Muller RH (1985a) In situ X-Ray diffraction of surface layers on lithium in nonaqueous electrolyte. J Electrochem Soc 132(6):1385–1387

Nazri G, Muller RH (1985b) Composition of surface layers on Li electrodes in PC, LiClO4 of very low water content. J Electrochem Soc 132(9):2050–2054

Nazri G, Muller RH (1985c) Effect of residual water in propylene carbonate on films formed on lithium. J Electrochem Soc 132(9):2054–2058

Neudecker BJ, Dudney NJ, Bates JB (2000) "Lithium-Free" thin-film battery with in situ plated Li anode. J Electrochem Soc 147(2):517–523

Nishida T, Nishikawa K, Rosso M, Fukunaka Y (2013) Optical observation of Li dendrite growth in ionic liquid. Electrochim Acta 100:333–341. doi:10.1016/j.electacta.2012.12.131

Nishikawa K, Fukunaka Y, Sakka T, Ogata YH, Selman JR (2007) Measurement of concentration profiles during electrodeposition of Li metal from LiPF6-PC electrolyte solution. J Electrochem Soc 154(10):A943–A948. doi:10.1149/1.2767404

Nishikawa K, Mori T, Nishida T, Fukunaka Y, Rosso M, Homma T (2010) In situ observation of dendrite growth of electrodeposited Li metal. J Electrochem Soc 157(11):A1212–A1217. doi:10.1149/1.3486468

Nishikawa K, Mori T, Nishida T, Fukunaka Y, Rosso M (2011) Li dendrite growth and Li$^+$ ionic mass transfer phenomenon. J Electroanal Chem 661(1):84–89. doi:10.1016/j.jelechem.2011.06.035

Nishikawa K, Naito H, Kawase M, Nishida T (2012) Morphological variation of electrodeposited Li in ionic liquid. ECS Trans 41:3–10

Odziemkowski M, Krell M, Irish DE (1992) A raman microprobe in situ and ex situ study of film formation at lithium/organic electrolyte interfaces. J Electrochem Soc 139(11):3052–3063

Orsini F, Du Pasquier A, Beaudoin B, Tarascon JM, Trentin M, Langenhuizen N, De Beer E, Notten P (1998) In situ scanning electron microscopy (SEM) observation of interfaces within plastic lithium batteries. J Power Sources 76:19–29

Orsini F, du Pasquier A, Beaudouin B, Tarascon JM, Trentin M, Langenhuizen N, de Beer E, Notten P (1999) In situ SEM study of the interfaces in plastic lithium cells. J Power Sources 81–82:918–921. doi:10.1016/S0378-7753(98)00241-9

Osaka T, Homma T, Momma T, Yarimizu H (1997a) In situ observation of lithium deposition processes in solid polymer and gel electrolytes. J Electroanal Chem 421:153–156

Osaka T, Momma T, Matsumoto Y, Uchida Y (1997b) Surface characterization of electrodeposited lithium anode with enhanced cycleability obtained by CO$_2$ addition. J Electrochem Soc 144(5):1709–1713

Ota H, Wang X, Yasukawa E (2004a) Characterization of lithium electrode in lithium imides/ethylene carbonate, and cyclic ether electrolytes. I. Surface morphology and lithium cycling efficiency. J Electrochem Soc 151(3):A427–A436. doi:10.1149/1.1644136

Ota H, Shima K, Ue M, Yamaki J-I (2004b) Effect of vinylene carbonate as additive to electrolyte for lithium metal anode. Electrochim Acta 49(4):565–572. doi:10.1016/j.electacta.2003.09.010

Ota H, Sakata Y, Wang X, Sasahara J, Yasukawa E (2004c) Characterization of lithium electrode in lithium imides/ethylene carbonate and cyclic ether electrolytes. II. Surface chemistry. J Electrochem Soc 151(3):A437–A446. doi:10.1149/1.1644137

Qian J, Xu W, Bhattacharya P, Engelhard M, Henderson WA, Zhang Y, Zhang J-G (2015a) Dendrite-free Li deposition using trace-amounts of water as an electrolyte additive. Nano Energy 15:135–144. doi:10.1016/j.nanoen.2015.04.009

Qian J, Henderson WA, Xu W, Bhattacharya P, Engelhard M, Borodin O, Zhang JG (2015b) High rate and stable cycling of lithium metal anode. Nat Commun 6:6362. doi:10.1038/ncomms7362

Rey I, Lassègues JC, Baudry P, Majastre H (1998a) Study of a lithium battery by confocal Raman microspectrometry. Electrochim Acta 43(10–11):1539–1544

Rey I, Bruneel J-L, Grondin J, Servant L, Lassègues J-C (1998b) Raman spectroelectrochemistry of a lithium/polymer electrolyte symmetric cell. J Electrochem Soc 145(9):3034–3042

Rosso M, Gobron T, Brissot C, Chazalviel JN, Lascaud S (2001) Onset of dendritic growth in lithium/polymer cells. J Power Sources 97–98:804–806

Rosso M, Brissot C, Teyssot A, Dollé M, Sannier L, Tarascon J-M, Bouchet R, Lascaud S (2006) Dendrite short-circuit and fuse effect on Li/polymer/Li cells. Electrochim Acta 51(25):5334–5340. doi:10.1016/j.electacta.2006.02.004

Ryou M-H, Lee DJ, Lee J-N, Lee YM, Park J-K, Choi JW (2012) Excellent cycle life of lithium-metal anodes in lithium-ion batteries with mussel-inspired polydopamine-coated separators. Adv Energy Mater 2(6):645–650. doi:10.1002/aenm.201100687

Sacci RL, Dudney NJ, More KL, Parent LR, Arslan I, Browning ND, Unocic RR (2014) Direct visualization of initial SEI morphology and growth kinetics during lithium deposition by in situ electrochemical transmission electron microscopy. Chem Commun 50(17):2104–2107. doi:10.1039/c3cc49029g

Sagane F, Ikeda K-I, Okita K, Sano H, Sakaebe H, Iriyama Y (2013) Effects of current densities on the lithium plating morphology at a lithium phosphorus oxynitride glass electrolyte/copper thin film interface. J Power Sources 233:34–42. doi:10.1016/j.jpowsour.2013.01.051

Sano H, Sakaebe H, Matsumoto H (2011) Observation of electrodeposited lithium by optical microscope in room temperature ionic liquid-based electrolyte. J Power Sources 196(16):6663–6669. doi:10.1016/j.jpowsour.2010.12.023

Sawada Y, Dougherty A, Gollub JP (1986) Dendritic and fractal patterns in electrolytic metal deposits. Phys Rev Lett 56(12):1260–1263. doi:10.1103/PhysRevLett.56.1260

Selim R, Bro P (1974) Some observations on rechargeable lithium electrodes in a propylene carbonate electrolyte. J Electrochem Soc 121(11):1457–1459

Shiraishi S, Kanamura K (1998) The observation of electrochemical dissolution of lithium metal using electrochemical quartz crystal microbalance and in-situ tapping mode atomic force microscopy. Langmuir 14:7082–7086

Shiraishi S, Kanamura K, Zi Takehara (1999) Surface condition changes in lithium metal deposited in nonaqueous electrolyte containing HF by dissolution-deposition cycles. J Electrochem Soc 146(5):1633–1639

Shiraishi S, Kanamura K, Takehara Z-I (2001) Imaging for uniformity of lithium metal surface using tapping mode-atomic force and surface potential microscopy. J Phys Chem B 105:123–134

Stark JK, Ding Y, Kohl PA (2011) Dendrite-free electrodeposition and reoxidation of lithium-sodium alloy for metal-anode battery. J Electrochem Soc 158(10):A1100–A1105. doi:10.1149/1.3622348

Stark JK, Ding Y, Kohl PA (2013) Nucleation of electrodeposited lithium metal: dendritic growth and the effect of co-deposited sodium. J Electrochem Soc 160(9):D337–D342. doi:10.1149/2. 028309jes

Steiger J (2015) Mechanisms of dendrite growth in lithium metal batteries. PhD Thesis, Karlsruhe Institute of Technology (KIT), Germany

Steiger J, Kramer D, Mönig R (2014a) Mechanisms of dendritic growth investigated by in situ light microscopy during electrodeposition and dissolution of lithium. J Power Sources 261:112–119. doi:10.1016/j.jpowsour.2014.03.029

Steiger J, Kramer D, Mönig R (2014b) Microscopic observations of the formation, growth and shrinkage of lithium moss during electrodeposition and dissolution. Electrochim Acta 136:529–536. doi:10.1016/j.electacta.2014.05.120

Steiger J, Richter G, Wenk M, Kramer D, Mönig R (2015) Comparison of the growth of lithium filaments and dendrites under different conditions. Electrochem Commun 50:11–14. doi:10. 1016/j.elecom.2014.11.002

Tang M, Albertus P, Newman J (2009) Two-dimensional modeling of lithium deposition during cell charging. J Electrochem Soc 156(5):A390–A399. doi:10.1149/1.3095513

Teyssot A, Belhomme C, Bouchet R, Rosso M, Lascaud S, Armand M (2005) Inter-electrode in situ concentration cartography in lithium/polymer electrolyte/lithium cells. J Electroanal Chem 584(1):70–74. doi:10.1016/j.jelechem.2005.01.037

Thompson RS, Schroeder DJ, López CM, Neuhold S, Vaughey JT (2011) Stabilization of lithium metal anodes using silane-based coatings. Electrochem Commun 13(12):1369–1372. doi:10. 1016/j.elecom.2011.08.012

Voss RF, Tomkiewicz M (1985) Computer simulation of dendritic electrodeposition. J Electrochem Soc 132(2):371–375

Wang X, Yasukawa E, Kasuya S (2000) Lithium imide electrolytes with two-oxygen-atom-containing cycloalkane solvents for 4 V lithium metal rechargeable batteries. J Electrochem Soc 147(7):2421–2426

Wenzel S, Leichtweiss T, Krüger D, Sann J, Janek J (2015) Interphase formation on lithium solid electrolytes—an in situ approach to study interfacial reactions by photoelectron spectroscopy. Solid State Ionics 278:98–105. doi:10.1016/j.ssi.2015.06.001

Yamaki J-I, Tobishima S-I, Hayashi K, Saito K, Nemoto Y, Arakawa M (1998) A consideration of the morphology of electrochemically deposited lithium in an organic electrolyte. J Power Sources 74:219–227

Yang L, Smith C, Patrissi C, Schumacher CR, Lucht BL (2008) Surface reactions and performance of non-aqueous electrolytes with lithium metal anodes. J Power Sources 185(2):1359–1366. doi:10.1016/j.jpowsour.2008.09.037

Yoon S, Lee J, Kim S-O, Sohn H-J (2008) Enhanced cyclability and surface characteristics of lithium batteries by Li–Mg co-deposition and addition of HF acid in electrolyte. Electrochim Acta 53(5):2501–2506. doi:10.1016/j.electacta.2007.10.019

Yoshimatsu I, Hirai T, Yamaki J-I (1988) Lithium electrode morphology during cycling in lithium cells. J Electrochem Soc 135(10):2422–2427

Zhamu A, Chen G, Liu C, Neff D, Fang Q, Yu Z, Xiong W, Wang Y, Wang X, Jang BZ (2012) Reviving rechargeable lithium metal batteries: enabling next-generation high-energy and high-power cells. Energy Environ Sci 5(2):5701–5707. doi:10.1039/c2ee02911a

Zhang Y, Qian J, Xu W, Russell SM, Chen X, Nasybulin E, Bhattacharya P, Engelhard MH, Mei D, Cao R, Ding F, Cresce AV, Xu K, Zhang JG (2014) Dendrite-free lithium deposition with self-aligned nanorod structure. Nano Lett 14(12):6889–6896. doi:10.1021/nl5039117

Zinigrad E, Aurbach D, Dan P (2001) Simulation of galvanostatic growth of polycrystalline Li deposits in rechargeable Li batteries. Electrochim Acta 46:1863–1869

Chapter 3
High Coulombic Efficiency of Lithium Plating/Stripping and Lithium Dendrite Prevention

As described in Chaps. 1 and 2, the Li plating morphology and the CE—both of which are critical for the safety characteristics and cyclability of a Li metal battery—are intricately linked with one another. Almost all the factors that lead to significant dendritic growth also lead to a lower CE and vice versa. This is because the needle-like ("dendritic") growth results in a greater amount of Li metal surface exposure to the electrolyte. This freshly formed Li metal surface is thermodynamically unstable against the electrolyte solvent molecules, ions, additives, and impurities with which it quickly reacts to form degradation products which may or may not fully passivate the Li surface from further reactions. This continuously consumes both the electrolyte and Li and thus leads to a lower CE. To have a high CE, the side reactions must be minimized. The proclivity for these to occur is proportional to the chemical and electrochemical activity of native Li with the neighboring electrolyte, as well as to the surface area of the deposited Li. Therefore, a high Li cycling CE may be achieved by the use of less reactive electrolyte components, a reduction in the exposed Li surface and the rapid passivation of this surface. The strong correlation between the Li plating/striping (or Li cycling) CE and Li morphology suggests that these should be discussed in tandem in Chap. 3 when considering the impact of differing Li cycling environments and conditions. Special cases of dendrite-free Li plating of highly aligned nanorod structures, but with a low CE (related to the high surface area of the Li nanorods) will also be discussed (Qian et al. 2015a; Ding et al. 2013b).

3.1 Coulombic Efficiency of Lithium Plating/Stripping

The scientific literature associated with Li plating/stripping frequently discusses the efficiency of this process, but the methods used to determine the CE vary and often result in significantly different values. The CE is often variable with cycling

© Springer International Publishing Switzerland 2017   45
J.-G. Zhang et al., *Lithium Metal Anodes and Rechargeable*
Lithium Metal Batteries, Springer Series in Materials Science 249,
DOI 10.1007/978-3-319-44054-5_3

conditions (current density, temperature) and may change considerably upon extended cycling or aging. Thus, reporting a specific value of the CE for a given electrolyte can be highly misleading without providing details of testing conditions.

Method 1 Using this method, the Li cycling CE is defined by Aurbach (1999):

$$CE = \frac{Q_S}{Q_D} \tag{3.1}$$

where Q_S and Q_D are the charges for Li stripping (dissolution) and plating (deposition) determined for each cycle, respectively. This method is often used when depositing Li on inert electrodes such as nickel (Ni), Cu, and tungsten (W).

Method 2 This method applies an initial known amount of deposition charge (Q_T) to plate Li metal on the working electrode (often Ni or Cu). Alternatively, if Li metal is used as the electrode, Q_T is determined from the electrode thickness and density. A fraction of this charge, i.e., the cycling charge (Q_D), is cycled across the cell for the stripping and subsequent plating with the overpotential for the process monitored during the cycling. When the overvoltage significantly increases (indicating depletion of all of the deposited Li or an increased impedance due to a highly resistive interphase), the test is completed. The CE from this method can therefore be determined by (Aurbach 1999)

$$CE_{avg} = \left[1 - \frac{Q_T}{NQ_D} \right] \times 100\% \tag{3.2}$$

where N is the total number of cycles for the cell (Appetecchi et al. 1998, 1999). In practice, it may take a long time to exhaust all Li (Q_T) deposited in the initial cycle. Therefore, a simplified formula for average CE in n cycles ($n < N$) has been used:

$$CE_{n-avg} = \frac{\left[Q_D - \frac{CE_{n-avg}Q_T - Q_r}{n} \right]}{Q_D} = \frac{nQ_D + Q_r}{nQ_D + Q_T} \times 100\% \tag{3.3}$$

where Q_r is the maximum amount of Li stripped from the working electrode after n cycles (Aurbach et al. 1989; Ding et al. 2013b). In the extreme case of $n = N$, Eq. (3.3) is equivalent to Eq. (3.2). For symmetric Li$\|$Li cells, Eq. (3.2) becomes (Appetecchi and Passerini 2002):

$$CE_{avg} = \left[1 - \frac{Q_T}{2NQ_D} \right] \times 100\% \tag{3.4}$$

The factor of two accounts for the two steps in each cycle in which the Li is reduced at one electrode and oxidized at the other. This effectively doubles the losses with respect to a battery in which the Li is only oxidized during the discharge and reduced during the charge (Appetecchi and Passerini 2002). This equation is derived from the single-step-efficiency equation:

$$CE = \frac{(Q_D - Q_L)}{Q_D} \times 100\%$$ (3.5)

where Q_L is the charge for the lost Li in each step (if $Q_L = 0$, the CE = 100 %). With Li electrode, however, Q_L can only be determined by cycling the cell until all of the charge for the Li is lost, i.e., $Q_T = 2nQ_L$ for a Li‖Li cell (at which point the overpotential increases rapidly). By substituting this into Eq. (3.5), the CE_{avg} Eq. (3.4) can be derived.

Method 3 Some reports use a figure-of-merit (FOM) instead of CE (Abraham and Goldman 1983) with

$$FOM_{Li} = \frac{nQ_D}{Q_T} = \frac{1}{1 - CE_{avg}}$$ (3.6)

The FOM_{Li} is thus [the total accumulated cell capacity]/[theoretical Li capacity or Li turnover number]. This value facilitates the recognition of differences in Li cyclability since an increase in the CE from 0.98 to 0.99 corresponds to an increase in the FOM from 50 to 100 which signifies a doubling of the cell cycle life (Arakawa et al. 1999).

3.2 Electrolyte and In Situ Formed Solid Electrolyte Interphase

The components of electrolytes used for Li^+ cation transport vary widely (Marcinek et al. 2015). Although ionic conductivity is most often used as a prime factor for electrolyte optimization since it is one of the determinants for the overall cell resistance and thus the power capability of a cell (Dudley et al. 1991), this property is of little utility for determining the stability of a given electrolyte with Li metal (Tobishima et al. 1995). The consequence of this is that the Li plating characteristics (i.e., morphology, reactivity, etc.) and their relationship to the plating/stripping current density are generally not directly correlated with variations in the electrolyte conductivity. Conditions that favor a high cycling CE include the following:

1. uniform (compact) Li deposition which minimizes the surface area of the plated Li exposed to the electrolyte,
2. the formation of a stable passivation layer which protects the Li from further reactions with electrolyte components—if such a layer is elastic, then it can stretch and contract as necessary to conform to changes in the Li volume and may aid in the redeposition of Li underneath such a layer thus reducing the exposure of fresh Li to the electrolyte (Aurbach 1999).

In particular, the deposition morphology is a function of the electrolyte solvent (s), salt(s), salt concentration, current density, temperature, and other factors. The SEI layer formed is often strongly influenced by contaminants or additives, even at the ppm level. Comparisons between the various studies reported in the literature must take this into consideration. In addition, in contrast with graphite anodes in which the passivating layer is supported on the graphite surface, the surface layers that form on electrodeposited Li may be, and usually are, disrupted when the Li is stripped away during cell discharge (for nodular deposits) or lack electrical contact with the current source (for needle-like deposits, which generate dead Li) (Fig. 2.17).

Although passivation films are known to form on the Li anode surface in batteries such as Li‖SOCl$_2$ primary cells, the role of this layer for Li stabilization was emphasized by Dey (1977) and (Peled 1979, 1983; Peled et al. 1997) who dubbed it a "solid electrolyte interphase" (SEI). The contributing components to the SEI and the effect that this has on Li plating/stripping will be discussed below. Note, however, that the study of the composition of the SEI layer is complicated by many factors, including the continuous evolution of the SEI during cycling/storage, changes that may occur from rinsing the surface (prior to ex situ characterization) or due to high vacuum conditions and transformations of the components due to Ar$^+$ sputtering during XPS depth analysis (Verma et al. 2010; Edström et al. 2006; Ota et al. 2004c). Thus, some caution should be exercised when interpreting the results reported in the diverse studies available in the literature.

Typically, as-received Li metal has surface coatings consisting principally of Li$_2$O (in contact with the Li) and an outer coating of LiOH and Li$_2$CO$_3$ (Hong et al. 2004; Kanamura et al. 1992, 1994b, 1995b, d, 1997; Shiraishi et al. 1995). Once the Li contacts the electrolyte, these native layers are transformed by both chemical and electrochemical reactions with the electrolyte components. The SEI films formed on Li in contact with an electrolyte are typically multi-layered with the species becoming more inorganic in nature (i.e., more fully reduced) the closer they are to the Li surface (Schechter et al. 1999). Over time, the interphasial layer may continue to grow in thickness due to continuous reactions, but the SEI composition may also be continuously transformed (Aurbach et al. 1994a, 1995b). Impurities such as dissolved gases (e.g., O$_2$, CO$_2$), H$_2$O, or others are often the most reactive constituents of the electrolyte and, as such, these reactions may dominate the characteristics of the as-formed SEI layers. For solvents that are relatively reactive (e.g., esters and carbonates), solvent degradation is the next most important factor in determining the composition of the SEI. If the electrolytes instead have solvents that are more stable to reduction (e.g., ethers), then anion degradation may be very influential in the SEI's composition (Aurbach et al. 1994b).

3.2.1 Influence of Solvents

Molecular orbital (MO) theory allows for an approximate estimation of the oxidative and reductive stability of a molecule or ion. A higher level of the highest occupied molecular orbital (HOMO) energy indicates that the species is a stronger electron donor and thus more susceptible to oxidation, while a lower level of the lowest unoccupied molecular orbital (LUMO) energy indicates a stronger electron acceptor, which is more readily reduced. The theory allows for an approximate estimation of the oxidative and reductive stability of a molecule or ion. Figure 3.1 indicates that according to MO theory ether solvents are significantly more stable to reduction than ester and carbonate solvents, but the ethers also tend to have a lower oxidative stability than carbonates (Wang et al. 1999). Note, however, that the stability of the solvents to oxidation and reduction is considerably different when they coordinate a cation (e.g., the formation of a coordinate bond involves electron

Fig. 3.1 Molecular orbital energies of aprotic solvents. Reproduced with permission —Copyright 1999, The Electrochemical Society (Wang et al. 1999)

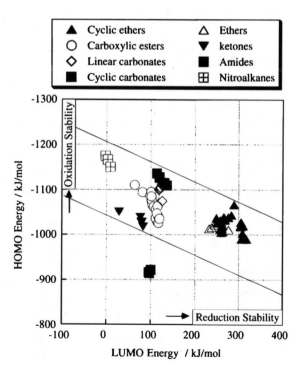

donation, which makes the solvent less/more susceptible to donating/accepting an electron to/from an electrode). For example, it has been demonstrated that the oxidative stability of glymes is greatly increased for highly concentrated electrolytes in which all of the solvent molecules are coordinated (Pappenfus et al. 2004; Yoshida et al. 2011). A corresponding decrease in reductive stability upon coordination is also expected (Aurbach and Gottlieb 1989).

Aurbach et al. (1997) used FTIR spectra (Fig. 2.11) to analyze the composition of SEI layers formed on Li electrodes prepared and stored for three days in EC-DMC solutions of 1 M LiAsF$_6$, LiPF$_6$ and LiBF$_4$. They have deduced and listed the possible components on the Li surface and their corresponding peaks in FTIR spectra (Aurbach et al. 1990a, 1994b, 1995c). Some possible chemical reactions on the Li surface have also been proposed as shown in Scheme 3.1) (Aurbach et al. 1990a, 1997).

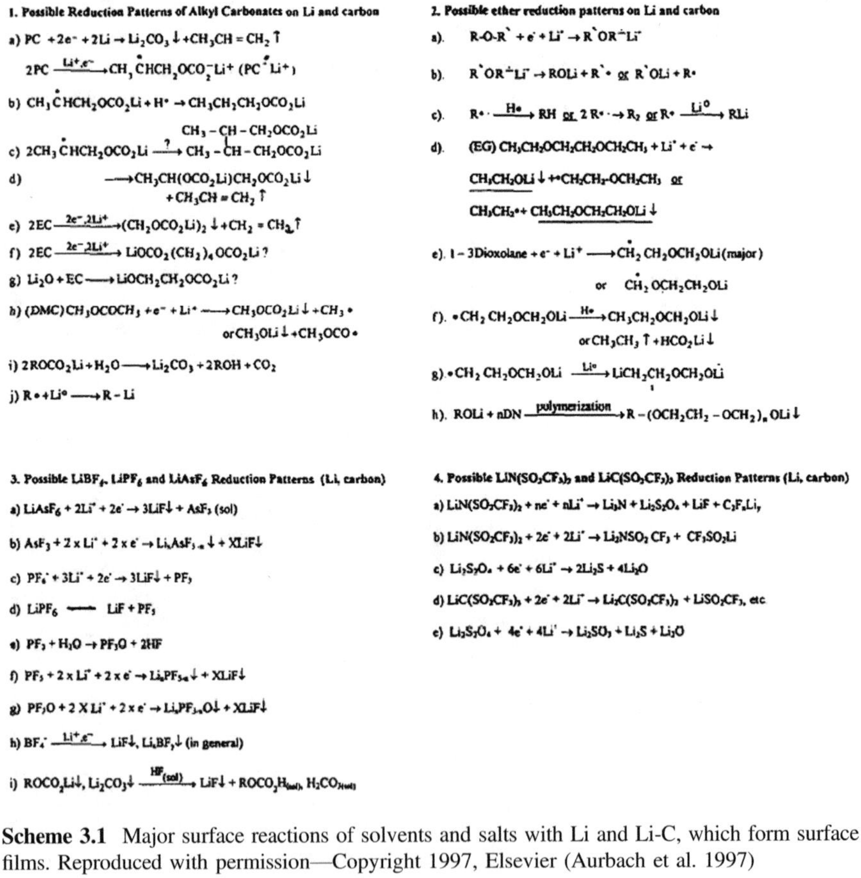

1. Possible Reduction Patterns of Alkyl Carbonates on Li and carbon

a) PC $+2e^- +2Li \rightarrow Li_2CO_3 \downarrow +CH_3CH=CH_2 \uparrow$

   $2PC \xrightarrow{Li^+,e^-} CH_3\overset{\bullet}{C}HCH_2OCO_2^-Li^+ (PC^{\bullet-}Li^+)$

b) $CH_3\overset{\bullet}{C}HCH_2OCO_2Li+H^\bullet \rightarrow CH_3CH_2CH_2OCO_2Li$

c) $2CH_3\overset{\bullet}{C}HCH_2OCO_2Li \xrightarrow{?} \begin{matrix} CH_3-CH-CH_2OCO_2Li \\ CH_3-CH-CH_2OCO_2Li \end{matrix}$

d) $\longrightarrow CH_3CH(OCO_2Li)CH_2OCO_2Li \downarrow$
   $+CH_3CH=CH_2 \uparrow$

e) $2EC \xrightarrow{2e^-,2Li^+} (CH_2OCO_2Li)_2 \downarrow +CH_2=CH_2 \uparrow$

f) $2EC \xrightarrow{2e^-,2Li^+} LiOCO_2(CH_2)_4OCO_2Li$?

g) $Li_2O + EC \longrightarrow LiOCH_2CH_2OCO_2Li$?

h) (DMC) $CH_3OCOCH_3 +e^- +Li^+ \longrightarrow CH_3OCO_2Li \downarrow +CH_3 \bullet$
   or $CH_3OLi \downarrow +CH_3OCO \bullet$

i) $2ROCO_2Li+H_2O \longrightarrow Li_2CO_3 +2ROH +CO_2$

j) $R \bullet +Li^\circ \longrightarrow R-Li$

2. Possible ether reduction patterns on Li and carbon

a). $R-O-R^\bullet +e^- +Li^+ \rightarrow R^\bullet OR^{\pm}Li^-$

b). $R^\bullet OR^{\pm}Li^- \rightarrow ROLi +R^\bullet \bullet$ or $R^\bullet OLi +R\bullet$

c). $R\bullet \xrightarrow{H\bullet} RH$ or $2R\bullet \rightarrow R_2$ or $R\bullet \xrightarrow{Li^\circ} RLi$

d). (EG) $CH_3CH_2OCH_2CH_2OCH_2CH_3 +Li^+ +e^- \rightarrow$
   $CH_3CH_2OLi \downarrow + \bullet CH_2CH_2-OCH_2CH_3$ or
   $CH_3CH_2\bullet + CH_3CH_2OCH_2CH_2OLi \downarrow$

e). $1-3$ Dioxolane $+e^- +Li^+ \longrightarrow \overset{\bullet}{C}H_2CH_2OCH_2OLi$ (major)
   or $\overset{\bullet}{C}H_2OCH_2CH_2OLi$

f). $\bullet CH_2CH_2OCH_2OLi \xrightarrow{H\bullet} CH_3CH_2OCH_2OLi \downarrow$
   or $CH_3CH_3 \uparrow +HCO_2Li \downarrow$

g). $\bullet CH_2CH_2OCH_2OLi \xrightarrow{Li\circ} LiCH_2CH_2OCH_2OLi$

h). $ROLi + nDN \xrightarrow{polymerization} R-(OCH_2CH_2-OCH_2)_n OLi \downarrow$

3. Possible LiBF$_4$, LiPF$_6$ and LiAsF$_6$ Reduction Patterns (Li, carbon)

a) $LiAsF_6 + 2Li^+ + 2e^- \rightarrow 3LiF\downarrow + AsF_3$ (sol)

b) $AsF_3 + 2 \times Li^+ + 2 \times e^- \rightarrow Li_xAsF_{3-x} \downarrow + XLiF\downarrow$

c) $PF_6^\bullet + 3Li^+ + 2e^- \rightarrow 3LiF\downarrow + PF_3$

d) $LiPF_6 \longleftarrow LiF + PF_5$

e) $PF_5 + H_2O \rightarrow PF_3O + 2HF$

f) $PF_5 + 2 \times Li^+ + 2 \times e^- \rightarrow Li_xPF_{5-x}\downarrow + XLiF\downarrow$

g) $PF_3O + 2 \times Li^+ + 2 \times e^- \rightarrow Li_xPF_{3-x}O\downarrow + XLiF\downarrow$

h) $BF_4^\bullet \xrightarrow{Li^+,e^-} LiF\downarrow, Li_xBF_y\downarrow$ (in general)

i) $ROCO_2Li\downarrow, Li_2CO_3\downarrow \xrightarrow{HF(sol)} LiF\downarrow + ROCO_2H_{(sol)}, H_2CO_{3(sol)}$

4. Possible LiN(SO$_2$CF$_3$)$_2$ and LiC(SO$_2$CF$_3$)$_3$ Reduction Patterns (Li, carbon)

a) $LiN(SO_2CF_3)_2 + ne^- + nLi^+ \rightarrow Li_3N + Li_2S_2O_4 + LiF + C_2F_xLi_y$

b) $LiN(SO_2CF_3)_2 + 2e^- + 2Li^+ \rightarrow Li_2NSO_2CF_3 + CF_3SO_2Li$

c) $Li_2S_2O_4 + 6e^- + 6Li^+ \rightarrow 2Li_2S + 4Li_2O$

d) $LiC(SO_2CF_3)_3 + 2e^- + 2Li^+ \rightarrow Li_2C(SO_2CF_3)_2 + LiSO_2CF_3$, etc.

e) $Li_2S_2O_4 + 4e^- + 4Li^+ \rightarrow Li_2SO_3 + Li_2S + Li_2O$

Scheme 3.1 Major surface reactions of solvents and salts with Li and Li-C, which form surface films. Reproduced with permission—Copyright 1997, Elsevier (Aurbach et al. 1997)

3.2.1.1   Esters

Esters (Fig. 3.2) are attractive electrolyte solvents due to their high oxidative stability and the very high conductivity of ester-based electrolytes, especially at low temperature. In addition to Li^+ cation solvation via the ester carbonyl oxygen, when methyl formate (MF) is used, anion solvation via hydrogen bonding to the MF formyl proton also occurs which further reduces ionic association interactions (Plichta et al. 1987; Venkatasetty 1975; Tobishima et al. 1990).

Rauh and Brummer reported that slow gas evolution occurs when Li is contacted with MF or methyl acetate (MA) and a white solid appears on the surface of the Li, whereas n-butyl formate (BF) was inert under the same conditions. $LiAsF_6$ addition improved the Li stability at elevated temperature, and the amount of gas evolution from Li immersed in $LiAsF_6$-ester electrolytes (heated up to 74 °C) indicated that the stability increased in the order MF > MA > ethyl acetate (EA) (Rauh and Brummer 1977b; Dampier and Brummer 1977). In general, the reactivity with Li increased with increasing alkyl chain length (methyl to butyl) and on transitioning from formates to acetates to propionates (Herr 1990). Tobishima et al., however, instead found that a 1.5 M $LiAsF_6$-MA electrolyte had a higher Li cycling CE than a 1.5 M $LiAsF_6$-MF electrolyte—which is also the case for the electrolytes with mixtures of these solvents and EC (Tobishima et al. 1989, 1990, 1995). The reason for this disagreement is unclear. Electrolytes with MA corrode Li to a much greater extent than those with PC, possibly due to the higher solubility of the reaction products resulting in the continuous exposure of Li to the solvent (Rauh and Brummer 1977b). It was found, however, that electrolytes in which MF or MA is mixed with a carbonate solvent such as DMC, DEC, or EC tend to have an improved CE (Tobishima et al. 1990; Plichta et al. 1989; Tachikawa 1993).

Herr (1990) indicated that esters may exist in both a stable keto form, as well as a reactive enol form. The enol form may readily be reduced by Li. Since MF does not

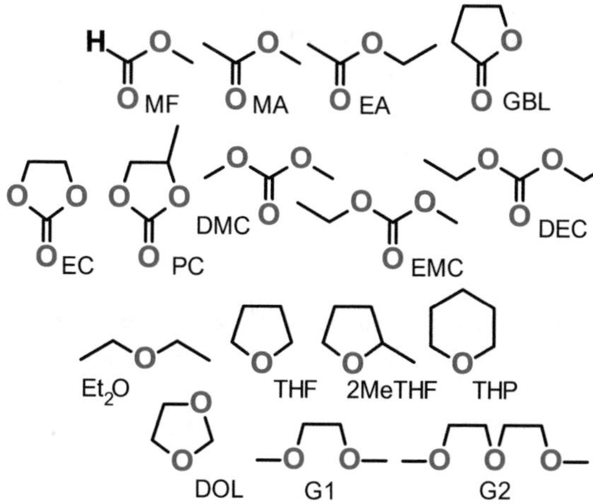

Fig. 3.2 Structures of selected aprotic solvents

form an enol, it cannot undergo such a reaction. Thus, MF is relatively stable to Li —in contrast to MA (Table 3.1.) (Herr 1990). Rauh (1975) indicated, however, that degradation may instead occur in a similar manner to the well-known reduction of carbonyl compounds by alkali metals. It is noteworthy that MF is susceptible to hydrolysis from trace amounts of H_2O (in both acidic and basic solutions) (Marlier et al. 2005), forming formic acid, which can then be dehydrated to form H_2O and CO; the H_2O from this then reacts with additional MF. Residual acidic $LiAsF_5OH$ in $LiAsF_6$ is believed to have catalyzed this reaction for MA-based electrolytes in early studies. The use of highly purified $LiAsF_6$ significantly decreased this reaction (Herr 1990). Hydrolysis can also occur in $LiBF_4$-MF electrolytes, but it was suggested that the LiF present with the $LiBF_4$ prevents the subsequent dehydration of the formic acid (Herr 1990). Thus, electrolytes with both $LiAsF_6$ and $LiBF_4$ in MF are more stable to Li than those without the latter salt as indicated by improved cycling performance and by an increase in the time (from 1 to 10 months on storage of Li in the electrolyte at 74 °C) before a detectable amount of gas generation was identified (Herr 1990; Plichta et al. 1989; Honeywell 1975; Ebner and Lin 1987). Instead of degrading to H_2O and CO, the formic acid may also directly react with the Li since spectroscopic evaluations of the SEI indicate that the major MF reduction product is Li formate (Table 3.1) (Aurbach and Chusid (Youngman) 1993; Ein-Eli and Aurbach 1996).

Cyclic esters, such as gamma-butyrolactone (GBL), have a high flash point and low vapor pressure which reduces the flammability of electrolytes with such solvents (Hess et al. 2015). The use of 1 M LiX-GBL electrolytes (with $LiAsF_6$, $LiClO_4$, $LiBF_4$, or $LiCF_3SO_3$) resulted in a relatively low Li cycling CE and the SEI formed in these electrolytes did not effectively passivate Li foil from continuous reactions on storage (Sazhin et al. 1994). This may be due to the relatively high solubility of the GBL reaction products (Aurbach and Gottlieb 1989; Aurbach et al. 1991a; Rendek et al. 2003; Aurbach 1989a, b). Kanamura et al. (1995d) noted that when Li foil was immersed in a 1 M $LiBF_4$-GBL electrolyte for 3 days, the electrode resistance initially decreased, but then increased steadily with time and the resulting SEI consisted predominantly of LiF with only a small amount of organic compounds present. The SEI formed on Li foil with GBL electrolytes with $LiAsF_6$, $LiPF_6$, or $LiClO_4$ also contained little to no organic components (Kanamura et al. 1995b). This may again be due to the high solubility of the organic salts produced from GBL degradation (Table 3.1).

3.2.1.2 Alkyl Carbonates

As is the case for esters, alkyl carbonates (Fig. 3.2) are not overly stable to Li reduction (Fig. 3.1). Cyclic carbonates (e.g., EC and PC) are either reduced to Li_2CO_3 and ethylene (or propylene) or the carbonyl group is instead reduced to form a radical anion which then further reacts to form organic salts and polymers

Table 3.1 Reaction products of solvents, salts, and contaminants with Li (Aurbach 1999)

Solvent	References	Dry	Contaminants/additives			
			H_2O	O_2	CO_2	HF
MF	Ein Ely and Aurbach (1992)	mostly HCO_2Li, ROLi (CH_3OLi)	HCO_2Li	Li oxides + HCO_2Li	$HCO_2Li + Li_2CO_3$	LiF
GBL	Aurbach and Gottlieb (1989), Aurbach (1989a)	$CH_3(CH_2)_2COOLi$, cyclic di-Li β-keto ester salt	$LiO(CH_2)_3CO_2Li$	Li oxides $ROCO_2Li$ species	$RCOOLi + Li_2CO_3$	LiF
DMC	Aurbach et al. (1994a)	$ROCO_2Li$ (CH_3OCO_2Li)	(Reaction of $ROCO_2Li + H_2O$)	Li_2O	Li_2CO_3	LiF+ $ROCO_2H$
DEC	Aurbach et al. (1987), (1995b)	$CH_3CH_2OCO_2Li + CH_3CH_2OLi$		Li_2O_2		
EMC	Ein-Eli et al. (1996)	CH_3OLi, CH_3OCO_2Li	$LiOH\text{-}Li_2O$			
PMC	Ein-Eli et al. (1997)	CH_3OLi, CH_3OCO_2Li, CH_3CH_2OLi, $CH_3CH_2OCO_2Li$,				
EC	Aurbach et al. (1992), (1994a)	$(CH_2OCO_2Li)_2$ + ethylene	Li_2CO_3			LiF+ H_2CO_3
PC	Aurbach et al. (1987), (1994a)	$CH_3CH(OCO_2Li)$ CH_2OCO_2Li + propylene	$ROCO_2Li$	$ROCO_2Li$	$ROCO_2Li$	LiF
THF	Aurbach et al. (1988)	ROLi, $CH_3(CH_2)_3OLi$	LiOH	Li oxides	Li_2CO_3	

(continued)

Table 3.1 (continued)

Solvent	References	Dry	Contaminants/additives			
			H_2O	O_2	CO_2	HF
2MeTHF	Malik et al. (1990)	Li pentoxides	Li_2O			
DOL	Aurbach et al. (1990a, b)	CH_3CH_2OLi $(CH_2OLi)_2$ poly DOL species			$ROCO_2Li$	
DME	Aurbach et al. (1988), Aurbach and Granot (1997)	ROLi (CH_3OLi)	Li alkoxides	ROLi species	ROLi species	
Et-DME	Aurbach and Granot (1997)	CH_3CH_2OLi $(CH_2OLi)_2$				
G2	Aurbach and Granot (1997)	CH_3OLi, $CH_3OCH_2CH_2OLi$, $(CH_2OLi)_2$				
THF-2MeTHF	Aurbach et al. (1995a)	THF reduction products dominate				
MF-DMC	Ein-Eli and Aurbach (1996)	HCO_2Li dominates, $ROCO_2Li$ (minor)	HCO_2Li, Li_2O-LiOH		Li_2CO_3 + red. products	LiF
MF-DEC	Ein-Eli and Aurbach (1996)		$LiO(CH_2)_3CO_2Li$		Li_2CO_3 + red. products	
MF-EC	Ein-Eli and Aurbach (1996)	HCO_2Li, $ROCO_2Li$ species	Li_2CO_3, HCO_3Li	$ROCO_2Li$	HCO_2Li	LiF
MF-PC	Ein-Eli and Aurbach (1996)		LiOH-Li_2O	HCO_2Li	Li_2CO_3	

(continued)

Table 3.1 (continued)

Solvent	References	Dry	Contaminants/additives			
			H_2O	O_2	CO_2	HF
MF-ethers	Ein-Eli and Aurbach (1996)	HCO_2Li dominates	Li_2CO_3, $ROCO_2Li$		Li_2CO_3 + red. products	
EC-DMC	Aurbach et al. (1996)					
EC-EMC	Ein-Eli et al. (1996)		$ROCO_2Li$	Li_2O	$ROCO_2Li$	
EC-DEC	Aurbach et al. (1995b)		Li_2O-LiOH			
EC-PC	Aurbach et al. (1992)	EC reduction products dominate	$ROCO_2Li$, Li_2CO_3	Li_2O, $ROCO_2Li$	Li_2CO_3, $ROCO_2Li$	LiF
EC or PC-ethers	Aurbach and Gofer (1991)	$ROCO_2Li$ species dominate, ROLi (minor)	LiOH-Li_2O			

(Gachot et al. 2008; Herr 1990; Nazri and Muller 1985a; Aurbach et al. 1987; Aurbach and Gottlieb 1989; Ota et al. 2004c; Wang et al. 2001). EC therefore is thus reduced principally to lithium ethylene dicarbonate $(CH_2OCO_2Li)_2$ and lithium ethoxide (CH_3CH_2OLi), as well as other degradation products such as $(CH_2CH_2OCO_2Li)_2$, $LiO(CH_2)_2CO_2(CH_2)_2OCO_2Li$, $Li(CH_2)_2OCO_2Li$, and Li_2CO_3 (Wang et al. 2001). PC undergoes similar reactions. PC (and EC) can also be nucleophilically attacked by OH^- from the native film (LiOH) or generated by the reaction of Li with trace H_2O yielding lithium alkyl carbonate(s) (e.g., $ROCO_2Li$) and Li_2CO_3 (Table 3.1) (Aurbach 1999).

Kanamura et al. noted that when Li foil was immersed in a 1 M $LiBF_4$-PC electrolyte for 3 days, the electrode resistance initially decreased, but then increased steadily with time, and the resulting SEI consisted predominantly of LiF with only a small amount of organic compounds present—as with GBL electrolytes—but the impedance was an order of magnitude higher than those for the comparable GBL electrolytes (Kanamura et al. 1995b). Koike et al. also found that a 1 M $LiClO_4$-PC electrolyte continuously reacted with the deposited Li, indicating that the SEI layer did not effectively passivate the Li (Koike et al. 1997; Nishikawa et al. 2010; Qian et al. 2015b; Ding et al. 2013a). Ding et al. (2013a) scrutinized the CE and morphology of Li deposited on Cu from 1 M $LiPF_6$-carbonate electrolytes (with PC, EC, DMC, and EMC). The deposits from cyclic carbonates (PC and EC) were clusters of thick, tangled needles, whereas those for the acyclic carbonates (DMC and EMC) were more fibrous and only partially covered the Cu substrate (Fig. 3.3). The surface layer of the SEI films from the PC and EC electrolytes were composed principally of LiF and Li_2CO_3, with more of the former for the PC electrolyte and more of the latter for the EC electrolyte. PC appears to be more reactive with the $LiPF_6$ salt since this electrolyte became discolored on storage, whereas the others did not. More organic compounds/salts that were less effective at passivating the Li surface were formed in the acyclic carbonates. The average Li cycling CE values for the 1 M $LiPF_6$-carbonate electrolytes were the following: PC (77 %), EC (95 %), DMC (24 %), and EMC (7 %) (Ding et al. 2013a).

Other studies have also shown that electrolytes with DMC generally result in a superior CE and surface passivation relative to those with DEC (or EMC) (Ding et al. 2013a; Tachikawa 1993). This despite the fact that lithium alkyl carbonate and lithium alkoxide reduction products are formed for all of these acyclic solvents (Table 3.1). Aurbach et al. (1987) noted that Li metal dissolves completely when immersed in pure DEC due to a lack of passivation, resulting in a brownish solution. Tachikawa (1993) also noted a brown discoloration of the solution during Li deposition (on a Ag electrode) using a 1 M $LiClO_4$-DEC electrolyte. Aurbach et al. (1987) reported that the main reaction products were lithium alkyl carbonates and lithium alkoxides (Table 3.1) with no detectable Li_2CO_3 evident. This suggests that the partially reduced alkyl carbonate undergoes radical reactions rather than a second electron transfer which would yield the fully reduced species (i.e., Li_2CO_3) (Gachot et al. 2008). This differs from the cyclic carbonates which do tend to form

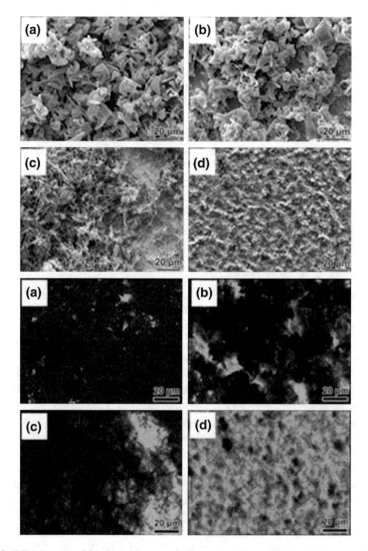

Fig. 3.3 SEM images of Li deposition morphologies and the corresponding energy-dispersive X-ray spectroscopy dot maps of Cu (represented by *green color*) from 1 M LiPF$_6$-carbonate electrolytes with **a** PC, **b** EC, **c** DMC, and **d** EMC. Reproduced with permission—Copyright 2013, The Electrochemical Society (Ding et al. 2013a)

Li$_2$CO$_3$, perhaps due to the greater stability of the intermediate radical for these cyclic solvents relative to that for DEC. The poor passivation and discoloration originate because the DEC reduction products—CH$_3$CH$_2$OLi and CH$_3$CH$_2$OCO$_2$Li —are soluble in DEC solutions (Schechter et al. 1999; Ding et al. 2013a; Tachikawa 1993).

3.2.1.3 Ethers

Ethers (Fig. 3.2) tend to be much more stable to reduction than esters and carbonates (Fig. 3.1) (Aurbach and Gottlieb 1989). Thus, the use of ethers as electrolyte solvents usually results in the highest CE when plating/stripping Li. Early work focused on diethyl ether Et_2O, THF, and 2MeTHF (Koch et al. 1982; Abraham et al. 1982). Kanamura et al. noted that when Li foil was immersed in a 1 M $LiBF_4$-THF electrolyte for 3 days, the electrode resistance variation with time differed markedly from that for electrolytes with GBL and PC. The resistance remained relatively low for approximately 36 h and then grew rapidly to a value comparable to that for the PC electrolyte. The resulting SEI consisted of both LiF and a relatively large amount of organic compounds. It was suggested that the HF present initially reacted with the native film to form LiF and the THF percolated through the native surface layer to the Li and reacted to form organic products (Kanamura et al. 1995d). $LiAsF_6$-2MeTHF electrolytes, however, result in a significantly higher Li cycling CE than do comparable $LiAsF_6$-THF electrolytes, as well as those with Et_2O (Goldman et al. 1980; Abraham and Goldman 1983; Abraham et al. 1982, 1986). It has been proposed that ring-opening reactions are significantly slower with 2MeTHF and this accounts for the improved cycling with this solvent (Goldman et al. 1980; Koch 1979).

Glymes behave somewhat differently from the cyclic ethers. Aurbach et al. proposed a mechanism for DME reduction in which a radical anion is formed which is stabilized by a Li^+ cation which decomposes to CH_3OLi and a methoxy ethyl radical. This then undergoes a second electron transfer producing an additional CH_3OLi and ethylene (Table 3.1) (Aurbach et al. 1988, 1993). Unfortunately, when used as single solvents for electrolytes with varying Li salts, glymes often result in crystalline solvates with high melting temperatures—except for electrolytes with dilute salt concentrations (Henderson 2006). There are thus few studies of Li plating/stripping in electrolytes with glyme solvents alone.

1,3-dioxolane (DOL) was found to be particularly useful as an electrolyte solvent. Dominey et al. reported that $LiClO_4$-DOL electrolytes resulted in smooth Li deposits with a very high Li cycling CE (Dominey and Goldman 1990; Shen et al. 1991; Dominey et al. 1991). Electrolytes with high purity were found to form SEI layers consisting of both alkoxides (e.g., $CH_3CH_2OCH_2OLi$) and LiCl (Aurbach et al. 1990b). $LiClO_4$-DOL electrolytes, however, were found to detonate due to an uncontrolled reaction as the electrolyte rapidly polymerized generating excessive heat (Newman et al. 1980). Replacement of the salt with $LiAsF_6$ and the use of additives such as 2mehtylfuran (2MF) and KOH—which scavenge the trace amounts of acidic species that polymerize DOL—resulted in a dramatic improvement in the cyclability of Li/TiS_2 cells Dominey and Goldman (1990; Goldman et al. 1989). Tributyl amine (Bu_3N) was also found to be a highly effective stabilizer to limit/prevent the chemical polymerization of the DOL for $LiAsF_6$-DOL electrolytes (Gofer et al. 1992). This formulation of electrolyte components resulted in nodular Li deposition with an exceptionally high Li cycling CE (Gofer et al. 1992;

Zaban et al. 1996; Dan et al. 1997; Aurbach and Moshkovich 1998; Zinigrad et al. 2004; Aurbach 1999; Aurbach et al. 2002b; Mengeritsky et al. 1996a).

3.2.1.4 Mixed Solvents

Interestingly, mixing two or more solvents is found, in some instances, to result in an improved Li cycling CE relative to comparable single solvent electrolytes. For example, this is the case for MF and acyclic carbonate (Table 3.1) (Salomon 1989; Tachikawa 1993; Plichta et al. 1989) and the addition of DMC or DEC to MF-based electrolytes was found to improve the cycling efficiency of Li||LiCoO$_2$ cells (Plichta et al. 1989). Single solvent electrolyte studies with EC have not been extensively studied due to this solvent's high melting temperature, but it has been widely used as a co-solvent. The combination of EC and other solvents increases the Li cycling CE, often quite significantly relative to the values for the single solvent electrolytes (including ethers), to average values >90 % (Fig. 3.4) (Tobishima et al. 1989, 1990). The Li surface chemistry of EC-DMC and EC-DEC mixtures is dominated by EC reduction, but this is more significant for Li passivation for the mixtures with DMC because the solvent reduction products are less soluble in DMC than in DEC (Schechter et al. 1999). In particular, Wang et al. found that electrolytes with EC, cyclic ethers, and lithium bis(perfluoroethylsulfonyl)imide (LiBETI) resulted in a good cycling performance, high thermal stability, high conductivity, and the homogeneous deposition of nodular Li with a relatively high CE (>90 %) (Wang et al. 1999, 2000; Xianming et al. 2001; Ota et al. 2004a). After 50 plating/stripping cycles, the interphasial layer was much thinner for the electrolyte with the cyclic ether (tetrahydropyranyl, THP) than for DME, carbonates (DMC and PC), and an ester (GBL) (Fig. 3.5), which accounts for the significant variances in the Li cycling performance with these electrolytes (Fig. 3.6) (Ota et al. 2004a). A detailed analysis of the SEI concluded that the outer layer consisted of ROCO$_2$Li (typical for EC), polymers, and LiF, while the inner layers contained Li$_2$O and carbide species with C–Li bonds (Ota et al. 2004c). The polymers were determined to be lithium

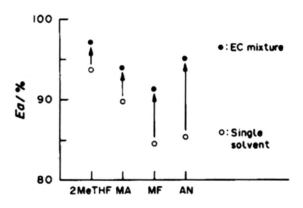

Fig. 3.4 Average Li cycling CE on a Pt substrate for 1.5 M LiAsF$_6$-solvent or LiAsF$_6$-EC/solvent (1:1) electrolytes. Reproduced with permission—Copyright 1990, Elsevier (Tobishima et al. 1990)

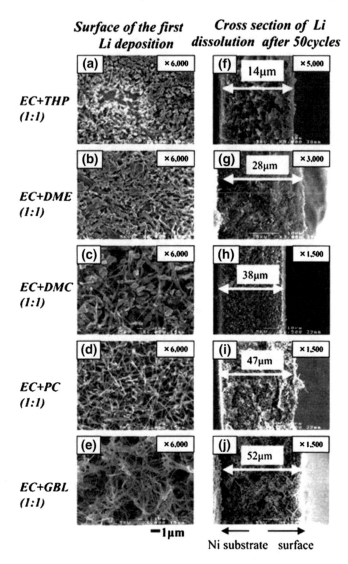

Fig. 3.5 SEM images of Li deposited on a Ni substrate in **a, f** EC/THP (1/1), **b, g** EC/DME (1/1), **c, h** EC/DMC (1/1), **d, i** EC/PC (1/1) and **e, j** EC/GBL (1/1) electrolytes containing 1 M LiBETI. The *left images* **a–e** are the surface morphologies after the initial plating at 0.6 mA cm^{-2} (0.5 C cm^{-2}). The *right images* **f–j** are the cross-sectional morphologies after 50 cycles at 0.6/0.6 mA cm^{-2} (0.5 C cm^{-2}) for the plating/stripping, respectively. Reproduced with permission—Copyright 2004, The Electrochemical Society (Ota et al. 2004a)

ethylene dicarbonate with ethoxy units suggesting that these polymers result from the reduction of both EC and the cyclic ether. Such polymers with both carbonate and flexible ethoxy segments may create a stable, elastic layer that more effectively

Fig. 3.6 Li cycling CE for the plating/stripping of Li on a Ni substrate in EC/THP (1/1), EC/DME (1/1), EC/DMC (1/1), EC/PC (1/1) and EC/GBL (1/1) electrolytes containing 1 M LiBETI. The current densities were 0.6/0.6 mA cm^{-2} (0.5 C cm^{-2}) for the plating/stripping, respectively. Reproduced with permission —Copyright 2004, The Electrochemical Society (Ota et al. 2004a)

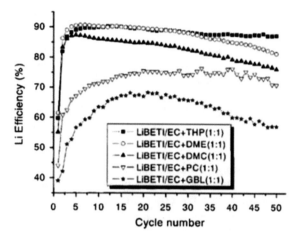

passivates the surface (than occurs for EC mixtures with glymes, carbonates, and esters), thus explaining the higher CE and reduced interfacial layer thickness.

3.2.2   Influence of Lithium Salts

Kanamura et al. (1995b) found that immersion of Li foil electrodes in 1 M LiX-GBL (with LiX is LiAsF$_6$, LiPF$_6$, LiClO$_4$, or LiBF$_4$) electrolytes resulted in only limited reactions of the native film with the LiAsF$_6$ and LiClO$_4$ electrolytes with major SEI species being LiOH and Li$_2$CO$_3$ and minor species of LiF (or LiCl) in the outer layer and Li$_2$O in the inner layer. For the LiBF$_4$ electrolyte, the SEI outer layer consisted principally of LiF with LiOH and Li$_2$CO$_3$ present as minor species and LiF and Li$_2$O as major and minor components, respectively, of the inner layer indicating the greater reactivity of the BF$_4^-$ anion. The SEI film formed in the LiPF$_6$ electrolyte, however, was thinner and more compact than that for the other electrolytes, but the resistance of this film was also much higher (with rapid reactions occurring just after immersion) (Kanamura et al. 1994b, 1995b). Koike et al. (1997) also reported that a 1 M LiPF$_6$-GBL electrolyte formed a thin, compact film on a Ni electrode during Li deposition. Kanamura et al. noted that the outer SEI layer resulting from the LiPF$_6$ electrolyte consisted principally of LiF with organic compounds, LiOH and Li$_2$CO$_3$ present as minor species, while the inner layer was composed of LiF and Li$_2$O. After 3 days of storage in the electrolytes, Li was plated onto the Li foil electrodes. This resulted in tangled dendritic needles for the LiAsF$_6$, LiClO$_4$, and LiBF$_4$ electrolytes, but the electrolyte with LiPF$_6$ seems to instead have formed a compact array of needles (Kanamura et al. 1994b, 1995b). This difference in behavior may perhaps be explained by the H$_2$O content in the electrolyte (reported as <100 ppm), since this amount of water is known to strongly influence HF formation and Li deposition morphology for a 1 M LiPF$_6$-PC electrolyte (Qian et al. 2015a).

Ding et al. (2013a) reported substantial differences in the Li morphology when plating from 1 M LiX-PC electrolytes. The average Li cycling CE values for the 1 M LiX-PC electrolytes were as follows:

$$\text{LiAsF}_6(95\%) \sim \text{LiBOB}(93\%) > \text{LiDFOB}(86\%) > \text{LiPF}_6(77\%)$$
$$> \text{LiCF}_3\text{SO}_3(73\%) \sim \text{LiClO}_4(72\%) \sim \text{LiTFSI}(72\%) \sim \text{LiBF}_4(72\%) \sim \text{LiI}(69\%)$$

Interestingly, the morphology reported for the lithium bis(oxalate)borate (LiBOB) electrolyte was highly fibrous, but resulted in a high CE, whereas the Li morphology for the electrolytes with the other salts resembled entangled thick needles or nodules, especially for the electrolytes that gave the highest CE values (Fig. 3.7) (Ding et al. 2013a). Aurbach has noted that a relatively high Li cycling CE is achievable, even for a fibrous Li deposition morphology, if the SEI contains a significant amount of Li$_2$CO$_3$ (Aurbach et al. 1992). This was attributed to the fact that the fibrous Li deposit formed with a Li$_2$CO$_3$ coating adhered very well to the

Fig. 3.7 SEM images of Li deposition morphologies from 1 M LiX-PC electrolytes with **a** LiBOB, **b** LiPF$_6$, **c** LiAsF$_6$, **d** LiTFSI, **e** LiI, **f** lithium difluoro(oxalato) borate (LiDFOB), **g** LiBF$_4$, **h** LiCF$_3$SO$_3$, and **i** LiClO$_4$. Reproduced with permission—Copyright 2013, The Electrochemical Society (Ding et al. 2013a)

electrode surface, in contrast to fibrous deposits formed in other electrolytes that lack Li_2CO_3 in the SEI layer. Although Li_2CO_3 might be expected to be a product from the reaction of the oxalate from the BOB^- anion, an XPS analysis found that Li_2CO_3 was absent. Alkyl oxalate and borate species may therefore instead account for the relatively high CE for the LiBOB electrolyte (Ding et al. 2013a).

For PC electrolytes with varying Li salts, Zaban and Aurbach (1995) differentiated between reactive salts ($LiBF_4$, $LiPF_6$, $LiCF_3SO_3$, and LiTFSI) and nonreactive salts (LiBr, $LiClO_4$, and $LiAsF_6$). When the nonreactive salts were used, the Li-electrolyte interface was stable upon storage, whereas the interfacial resistance continuously increased upon storage for electrolytes with reactive salts and was strongly dependent upon salt concentration. Geronov et al. (1982), however, reported that the Li-electrolyte resistance for 0.5 M LiX-PC electrolytes did increase over time, but to a much lesser extent for the electrolyte with $LiAsF_6$ than for the one with $LiClO_4$. Kanamura et al. (1997) also found that the SEI formed on Li immersed in a 1 M $LiClO_4$-DEC electrolyte became thick and was composed of a mixture of LiOH, Li_2CO_3, alkyl carbonates, alkoxides, and Li_2O, whereas immersion of Li in a 1 M $LiPF_6$-DEC electrolyte resulted in a compact SEI composed mostly of an inner layer of Li_2O and an outer layer of LiF (Figs. 3.8 and 3.9).

Wang et al. (1999) noted that for 1 M LiX-EC/THF (1/1) electrolytes, the Li cycling CE increased in the order as follows:

$$LiTFSI \approx LiBETI > LiClO_4 > LiBF_4 > LiPF_6$$

These authors found that the SEI formed during the plating/stripping from the $LiPF_6$ electrolyte contained a significant amount of LiF and residual Li was present after the stripping step, whereas very little LiF was present for the LiTFSI electrolyte and no residual Li was found after the stripping step. These authors also reported an improved CE for LiTFSI- and LiBETI-based electrolytes relative to $LiPF_6$-based electrolytes with EC/DME and EC/DEC mixed solvents (Wang et al. 2000a). Zaban et al. (1996) examined the resistance of the Li-electrolyte interface upon storage in 1 M LiX-EC/DEC (3/1) electrolytes. Notably, the resistance of the interface was an order of magnitude higher for the electrolytes with $LiPF_6$ and $LiBF_4$ relative to those with $LiAsF_6$ and $LiClO_4$ (Figs. 3.10 and 3.11).

Tobishima et al. (1995) found that the CE for 1 M LiX-EC/2MeTHF (1:1) electrolytes increased in the following order:

$$LiAsF_6 > LiClO_4 > LiBF_4 > LiSbF_6 > LiCF_3SO_3 > LiPF_6$$

Herr (1990) reported that the CE for 1 M LiX-MF electrolytes increased in the following order:

$$LiAsF_6 > LiClO_4 > LiPF_6 > LiBF_4$$

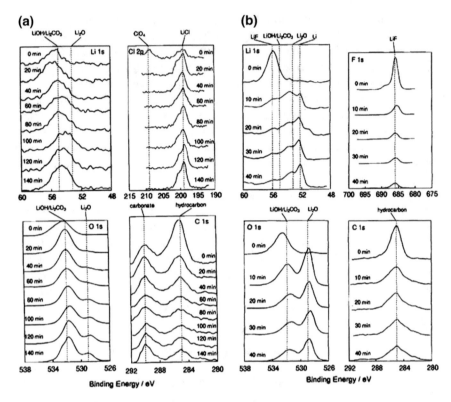

Fig. 3.8 XPS spectra for Li foil immersed in 1 M LiX-DEC for 240 min: **a** LiClO$_4$ and **b** LiPF$_6$. Reproduced with permission—Copyright 1997, The Electrochemical Society (Kanamura et al. 1997)

Fig. 3.9 Schematic model for the surface products on Li foil immersed in 1 M LiX-DEC: **a** LiClO$_4$ and **b** LiPF$_6$. Reproduced with permission—Copyright 1997, The Electrochemical Society (Kanamura et al. 1997)

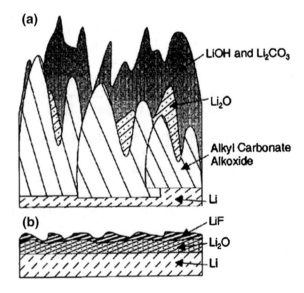

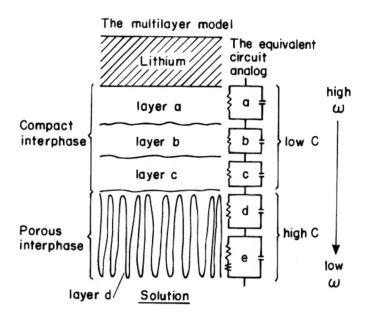

Fig. 3.10 Multilayer model of the Li-electrolyte interface and the equivalent circuit analog fit. Layers **a–c** may be thought of as the SEI, while the porous interphase is the interphasial layer which includes the dead or mossy Li. Reproduced with permission—Copyright 1996, The American Chemical Society (Zaban et al. 1996)

while the values for 1 M LiX-PC electrolytes increased in the following order (Herr 1990):

$$LiAsF_6 > LiClO_4 > LiBF_4 \sim LiPF_6.$$

The electrolyte salt thus plays a prominent role in the Li cycling performance. The strong disparity in behavior noted for $LiAsF_6$ and $LiPF_6$, despite their structural similarities, may be due to the substantial differences in the reductive stability of the anions. A study of THF and sulfolane electrolytes with $LiAsF_6$ and $LiPF_6$ found that AsF_6^- tend to be reduced to AsF_3 as follows:

$$AsF_6^- + 2e^- \leftrightarrow AsF_3 + 3F^-$$

$$F^- + Li^+ \leftrightarrow LiF$$

The LiF is insoluble in the electrolyte and thus remains on the electrode surface (i.e., becomes part of the SEI). In contrast, the PF_6^- anion is not readily reduced, but this anion does disproportionate as follows:

$$PF_6^- \leftrightarrow PF_5 + F^-$$

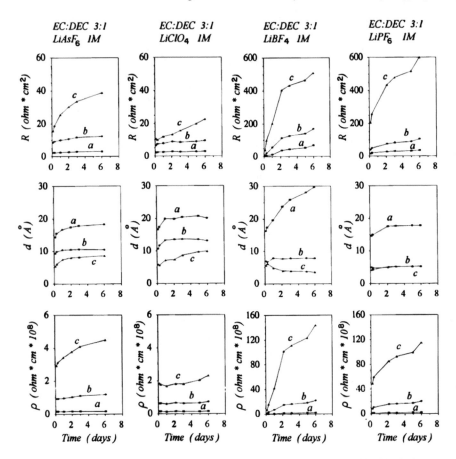

Fig. 3.11 Typical resistance, thickness, and resistivity for layers **a–c** (model in Fig. 3.10) as a function of storage time for 1 M LiX-EC/DEC (3/1) electrolytes. Reproduced with permission—Copyright 1996, The American Chemical Society (Zaban et al. 1996)

The PF_5 can then initiate the polymerization of cyclic ethers, while the F^- forms LiF (Nanjundiah et al. 1988). However, unlike the SEI formed with $LiPF_6$-based electrolytes which contain little phosphorus, those produced from $LiAsF_6$-based electrolyte contain arsenic and arsenic compounds (Aurbach et al. 2002a; Zaban et al. 1996):

$$AsF_6^- + 2Li \rightarrow 2LiF + AsF_3$$

$$AsF_3 + xLi \rightarrow Li_xAsF_{3-x} + xLiF$$

These arsenic compounds are believed to contribute to the favorable SEI characteristics that result in greatly improved Li cycling performance for $LiAsF_6$ electrolytes relative to comparable electrolytes with other Li salts.

3.2.3 Influence of Additives

A diverse range of additives have been added to electrolytes—either intentionally or accidentally as contaminants—which are often the most reactive components of the electrolyte and thus dominate the initial SEI formation even when present only in trace amounts. The most commonly studied contaminants are H_2O and O_2, while CO_2 has received intense focus as an additive.

3.2.3.1 H₂O and HF

H_2O is a frequent electrolyte contaminant. As such, many studies have been devoted to scrutinizing its influence on the Li plating morphology and cycling behavior. Small amounts of H_2O are often found to improve the Li cycling CE, but an excess amount of H_2O instead degrades the Li cycling CE (Malik et al. 1990; Rauh and Brummer 1977b; Aurbach and Granot 1997; Qian et al. 2015a; Selim and Bro 1974; Rauh and Brummer 1977a; Togasaki et al. 2014; Arakawa et al. 1999). When H_2O is present in carbonate-based electrolytes, its reaction suppresses the solvent and anion reduction reactions due to the formation of Li_2CO_3 (Schechter et al. 1999; Aurbach et al. 1995b; Zaban et al. 1996; Nazri and Muller 1985c):

$$2ROCO_2Li + H_2O \rightarrow 2ROH + CO_2 + Li_2CO_3$$

The alcohol produced is likely to be soluble in the electrolyte. The dissolved CO_2 may react with Li_2O or LiOH to form Li_2CO_3, or alternatively with a lithium alkoxide to form a lithium alkylcarbonate species. The Li_2CO_3 is insoluble in aprotic solvents and thus becomes part of the SEI layer. In ether-based electrolytes, however, H_2O does not react with the alkoxides formed, but it does solvate these surface species which facilitates the diffusion of the H_2O to the Li surface to form LiOH and Li_2O accompanied by H_2 generation (Zaban et al. 1996; Nazri and Muller 1985c; Malik et al. 1990):

$$2H_2O + 2Li \rightarrow 2LiOH + H_2$$

$$2LiOH + 2Li \rightarrow 2Li_2O + H_2$$

Therefore, small amounts of H_2O tend to increase the Li cycling CE for PC-based electrolytes, but decreases the CE for ether-based electrolytes (Aurbach et al. 1989).

Although the ClO_4^- anion is relatively stable to hydrolysis, this is not the case for PF_6^- and BF_4^- (Kanamura et al. 1995b, d; Qian et al. 2015a):

$$MF_n^- + H_2O \rightarrow F^- + 2HF + MOF_{n-3} \ (M = P \text{ or } B)$$

The F^- can then react with a Li^+ cation to form LiF, while the HF can react with several other compounds present in electrolyte or SEI layer according to the following reactions:

$$LiOH + HF \rightarrow LiF + H_2O$$
$$Li_2O + 2HF \rightarrow 2LiF + H_2O$$
$$2Li + 2HF \rightarrow 2LiF + H_2$$
$$ROCO_2Li + HF \rightarrow LiF + ROCO_2H$$
$$Li_2CO_3 + HF \rightarrow LiF + LiHCO_3$$
$$LiHCO_3 + HF \rightarrow LiF + H_2CO_3$$

The $ROCO_2H$ and H_2CO_3 are likely to be soluble in the electrolyte. Thus, electrolytes with $LiPF_6$, especially when a small amount of H_2O is present, often have a compact SEI layer consisting predominantly of LiF with relatively little organic species present. It is important to note that this does not necessarily mean that reactions with the solvent are largely prevented (although this may occur if the surface passivation is effective) since the products of such solvent reactions may continuously react with HF to form soluble species (which may then influence the cathode reactions). This LiF-dominant SEI layer on Li metal results in the uniform formation of Li nuclei underneath this inorganic layer which then may grow into an aligned array of Li nanorods (which is covered by an SEI layer consisting of LiF and perhaps some Li_2O) (Fig. 2.16) with a flat, mirror-like surface that results in a blue texture-coloring (Qian et al. 2015a).

Kanamura and Takehara reported that the direct addition of HF (e.g., to a 1 M $LiClO_4$-PC electrolyte) resulted in Li (plated on Ni) with a smooth, uniform deposition morphology that also had a blue color (Kanamura et al. 1994a; Takehara 1997). This is due to the formation of a thin surface coating of LiF (Kanamura et al. 1994a, 1995a, 1996, 2000; Takehara 1997; Shiraishi et al. 1997; Lee et al. 2005). The same results were obtained with electrolytes having different carbonate solvents and Li salts (Shiraishi et al. 1997). These authors indicated that the Li had a hemispherical form, but an examination of the SEM images suggests that this corresponds to the rounded tops of aligned arrays of Li nanorods (Kanamura et al. 1994a, 1995a; Shiraishi et al. 1997, 1999a; Yoon et al. 2008), as was noted above. Despite this favorable Li deposition morphology, the Li cycling CE in the 1 M $LiClO_4$-PC electrolyte with added HF was rather low (<90 %) (Kanamura et al. 2000; Shiraishi and Kanamura 1998; Shiraishi et al. 1999a; Yoon et al. 2008). This low efficiency may be attributed to a coating on the surface of each of the nanorods growing outward. After stripping the Li, a thick residual film remained on the electrode. This negatively impacted the subsequent depositions due to the accumulation of this interphasial layer (Shiraishi et al. 1999b; Yoon et al. 2008).

3.2.3.2 O_2 and N_2

The addition of dissolved O_2 tended to improve the Li cycling efficiency (Malik et al. 1990; Koch and Young 1978). In aprotic electrolytes, the reduction of O_2 (above 1.8 V vs. Li/Li$^+$—Fig. 3.12) results in a metastable superoxide ion or a peroxide (Pletcher et al. 1994; Aurbach et al. 1991b; Aurbach 1989a)

$$O_2 + Li \rightarrow LiO_2$$
$$O_2 + 2Li \rightarrow Li_2O_2$$

According to the recent Li-air battery studies, LiO_2 is reactive to carbonate solvents to Li_2CO_3 and lithium alkylcarbonate, while it is relatively stable in ethers and has certain solubility. These oxides should be highly insoluble in carbonate and ether solvents.

The reactivity of dissolved N_2 toward Li at low potential was studied in N_2-saturated 2MeTHF solutions (Malik et al. 1990). The reactivity to form Li_3N was rather low suggesting that N_2 is probably not reactive enough to compete with other solution components for the SEI formation. Only modest improvements to the CE were noted for an N_2-saturated 1 M LiAsF$_6$-THF electrolyte (Koch and Young 1978). Another study in which Li was exposed to dried air (i.e., a dry room), however, determined that the surface coating consisted of both Li_2CO_3 and Li_3N which improved the Li cycling CE relative to Li stored in Ar (Momma et al. 2011). Exposing Li directly to N_2 gas also produced a Li_3N layer which, when cycled in a 1 M LiPF$_6$-EC/DMC (1/1) electrolyte, resulted in stable cycling performance, suggesting that the Li_3N layer was durable, but the Li cycling CE was low (<90 %) for this electrolyte (Wu et al. 2011).

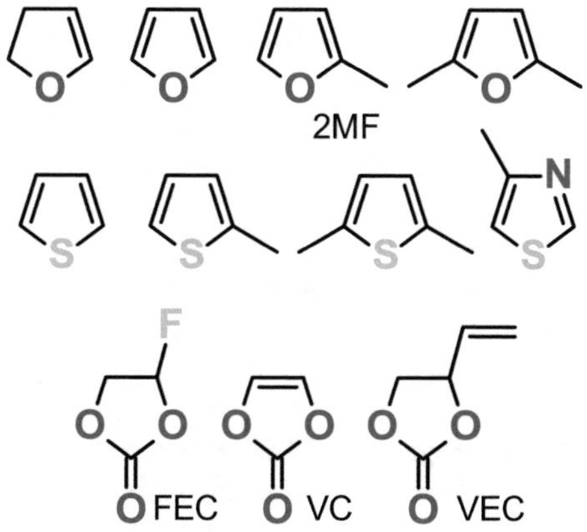

Fig. 3.12 Structures of selected organic additives

3.2.3.3 CO_2 and SO_2

Brummer et al. noted in 1977 that the addition of SO_2 improved the Li cycling CE of electrolytes (Rauh and Brummer 1977a, b; Dampier and Brummer 1977). This additive, however, did not effectively passivate the Li surface and its toxicity has limited its use. The use of CO_2, however, has been extensively studied as an additive. This reacts with surface species to form Li_2CO_3 which becomes a major surface species which, in turn, often greatly improves the Li cycling CE (Ein-Eli and Aurbach 1996; Aurbach and Chusid (Youngman) 1993; Ein Ely and Aurbach 1992; Malik et al. 1990; Osaka et al. 1997b, c; Besenhard et al. 1993; Fujieda et al. 1994; Momma et al. 1995; Gan and Takeuchi 1996). Li_2CO_3 is one of the most effective Li passivating species for preventing Li corrosion and the formation of dead Li (Besenhard and Eichinger 1976). Although the Li deposition morphology remains that of kinked needles, these adhere well to the electrode surface thus preventing the loss of electrical contact (Aurbach et al. 1992). Note, however, that this plating/stripping behavior will still result in the continuous buildup of an interphasial layer due to the residue from the needle-like SEI layers once the Li is stripped and then re-plated since the Li does not redeposit in the empty needle husks. In addition, excessive amounts of CO_2 can degrade the CE, especially for $LiAsF_6$-ether electrolytes, which tend to generate a stable SEI layer, resulting in a very high Li cycling CE (Malik et al. 1990; Koch and Young 1978). Of note, Aurbach et al. (1992) found that the use of Al_2O_3 as a drying agent for EC or PC produced CO_2 and thus storage of such electrolytes over Al_2O_3 will increase the Li cycling CE considerably.

3.2.3.4   Other Additives

Numerous other additives have been utilized. These include nitrous oxides (e.g., N_2O and CH_3NO_2) (Dampier and Brummer 1977; Rauh and Brummer 1977a, b; Uchiyama et al. 1987; Besenhard et al. 1993), trialkylamines (Aurbach et al. 2002a; Dan et al. 1997; Gofer et al. 1992; Herlem et al. 1996), aromatic compounds (e.g., benzene, toluene, and furans/thiophenes (especially 2MF) (see Fig. 3.12) (Abraham 1985; Morita et al. 1992; Geronov et al. 1989; Goldman et al. 1989; Halpert et al. 1994; Malik et al. 1990; Matsuda 1993; Abraham et al. 1984; Dominey and Goldman 1990; Ishikawa et al. 1994; Matsuda and Morita 1989; Matsuda and Sekiya 1999; Surampudi et al. 1993; Wu et al. 2013; Saito et al. 1997; Choi et al. 2007a; Peled et al. 1989; Choi et al. 2008; Lane et al. 2010a), and reactive carbonates (fluoroethylene carbonate, FEC and vinylene carbonate, VC) (Song et al. 2013a; Mogi et al. 2002c; Roberts et al. 2014; Guo et al. 2015; Ota et al. 2004b, d). FEC, VC, and vinylethylene carbonate (VEC) (see Fig. 3.12) have also been used as single solvents for 1 M $LiPF_6$-carbonate electrolytes which resulted in substantial differences in the Li morphology (Fig. 3.13) and comparable CE values of 97–98 % (compare with Fig. 3.3) (Ding et al. 2013a).

Fig. 3.13 SEM images of Li deposition morphologies from 1 M $LiPF_6$-carbonate electrolytes with **a** VC, **b** FEC, and **c** VEC. Reproduced with permission —Copyright 2013, The Electrochemical Society (Ding et al. 2013a)

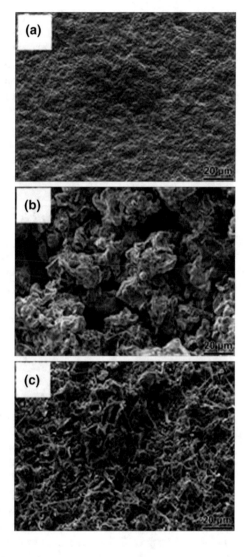

Salts have also been used as additives. These include salts with active cations such as multivalent cations (Mg^{2+}, Ca^{2+}, Zn^{2+}, Fe^{2+}, In^{3+}, Bi^{3+}, Al^{3+}, Ga^{3+}, Sn^{4+}) (Matsuda 1993; Aurbach et al. 2002b; Yoon et al. 2008), which are believed to form intermetallic alloy phases with Li at the electrode surface, as well as salts with active monovalent cations such as Na^+ (Gorodyskii et al. 1989; Bale 1989c; Stark et al. 2011, 2013). It is interesting to note that the phase diagram for Li–Na indicates that solid solutions are formed for all compositions that remain solid up until 92 °C, whereas the phase diagram for Li–Cs shows that these elements have almost complete immiscibility with each other in both the solid and liquid states and that Cs melts at 28 °C (Bale 1989b, c). Lithium halides (i.e., LiF, LiBr, and LiI)

have also been used as additives (Lu et al. 2014, 2015; Ishikawa et al. 1994; Besenhard 1977). Both LiF and LiBr have a low to negligible solubility in aprotic electrolytes, but LiI has an appreciable solubility. It was suggested, however, that the I^- anion absorbs on the Li surface. These halide salts cannot be further reduced, and so their presence on the Li surface may hinder or prevent reaction of the Li with other electrolyte components. Li salts with reactive anions (LiBOB, LiDFOB, and LiNO$_3$) have also been used (Xiong et al. 2011; Wu et al. 2014; Guo et al. 2015; Li et al. 2015; Zhang 2013a), as have salts that combine these features (MgI$_2$, AlI$_3$, and SnI$_2$) (Ishikawa et al. 1994; Matsuda et al. 1995; Ishikawa et al. 1997, 1999b, 2003, 2005).

3.2.4 Influence of Ionic Liquids

Many ionic liquids (ILs) with tetraalkylammonium (e.g., N^+_{RRRR}, PY^+_{1R}, and PI^+_{1R}) and tetraphosphonium (P^+_{RRRR}) cations have very high electrochemical stability windows. For example, for the N_{1116}TFSI IL (with trimethylhexylammonium cations), this is in excess of 5 V (0–5 V vs. Li/Li$^+$) when cycled with electrodes such as Pt, Al, or glassy carbon (Zheng et al. 2005). Notably, however, the electrochemical stability is strongly dependent on the electrode material as oxidation/reduction reactions were evident within this voltage range when Ni, Cu, and acetylene black were used instead (Zheng et al. 2005). Typically, the reductive stability of an IL is determined by the identity of the organic cation. ILs with 1-alkyl-3-methylimidazolium (e.g., $IM_{102}{}^+$) cations exhibit poor reductive stability due to the high reactivity of the C-2 proton on the cation's ring (Best et al. 2010; Matsumoto et al. 2006; Sano et al. 2013). It thus follows that ILs with 1-alkyl-2,3-dimethylimidazium (e.g., $IM_{113}{}^+$) cations are more stable to reduction (although the C-4 and C-5 protons are still somewhat acidic), while tetraalkylammonium and tetraalkylphosphonium cations are the most stable (Best et al. 2010; Matsumoto et al. 2006; Seki et al. 2006b).

In general, neat tetraalkylammonium-based ILs display a reductive current in its cyclic voltammetry curve just prior to the potential at which Li plating initiates. This would suggest that these ILs are unsuitable for this application (Bhatt et al. 2010), but when a Li salt (e.g., LiBF$_4$, LiPF$_6$, and LiTFSI) is dissolved in an IL (with either FSI$^-$ or TFSI$^-$ anions), a substantial extension of the cathodic stability limit occurs which enables the highly reversible plating/stripping of Li metal (Best et al. 2010; Matsumoto et al. 2006). The stabilization effects for the different Li salts were found to be comparable, irrespective of the Li salt used (Girard et al. 2015; Liu et al. 2014a; Best et al. 2010). The influence of temperature on electrochemical stability is often neglected. Few available studies explore this, but one study found that as the temperature increased, the reductive stability potential range of the electrolyte contracted resulting in the overlap of the Li plating process with electrolyte degradation and thus significant IL decomposition occurred at $\sim$ 80 °C (Best et al. 2010).

Howlett et al. (2004) reported that a Li cycling CE > 99 % could be achieved on a Cu electrode using a 0.5 M LiTFSI-IL electrolyte with the PY_{14}TFSI IL up to a current density of 1.5 mA cm^{-2}, but these measurements used an excess of initially plated Li, so the value per cycle may actually be somewhat lower. At higher current densities, kinked needle-like Li deposits formed. Nishida et al. (2013) explored the Li deposition morphology on a Ni electrode in LiTFSI-IL electrolytes with the $PY_{1(101)}$TFSI IL. These authors also studied the effect of LiTFSI salt concentration (Fig. 3.14). For the electrolytes with a LiTFSI concentration <1 M, aggregated dendrite clusters formed, but for more concentrated electrolytes, longer kinked needle-like deposits were evident.

A range of studies have been reported for Li anodes with electrolytes having ILs consisting of tetralkylammonium or tetraalkylphosphonium cations and either TFSI$^-$ or FSI$^-$ anions mixed with Li salts (LiTFSI, LiFSI, $LiAsF_6$, $LiPF_6$, $LiBF_4$) (Sakaebe and Matsumoto 2003; Howlett et al. 2004; Seki et al. 2005, 2006a, 2008; Matsumoto et al. 2006; Fernicola et al. 2007; Vega et al. 2009; Wibowo et al. 2009; Best et al. 2010; Bhatt et al. 2010; Lane et al. 2010b; Sano et al. 2011a, b, 2012, 2013, 2014 Basile et al. 2013; Goodman and Kohl 2014; Liu et al. 2014a; Furuya et al. 2015; Girard et al. 2015). Since no solvent is present in these electrolytes, the anion plays a prominent role in the SEI formation. For example, for Li stored in a LiFSI-IL electrolyte with PY_{13}FSI, the passivation layer contained LiF, Li_2O,

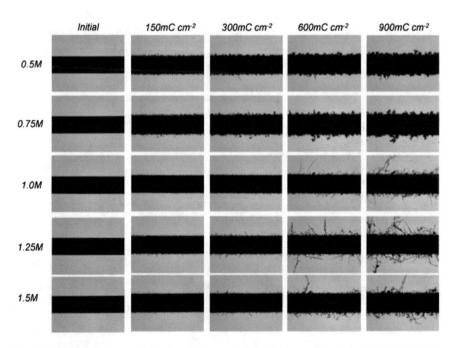

Fig. 3.14 Time transients of the morphology of plated Li on a Ni substrate in a LiTFSI-IL electrolyte as a function of LiTFSI salt concentration and total current. Reproduced with permission—Copyright 2013, Elsevier (Nishida et al. 2013)

LiOH, and FSI⁻ decomposition products (Budi et al. 2012). A similar complex mixer of species was proposed for the degradation of the TFSI⁻ anion (Howlett et al. 2004). But a significant difference was noted when the cell was cycled as the SEI then did not contain the species present in the native Li SEI layer (i.e., Li_2O, LiOH, and Li_2CO_3), but instead had significant quantities of species associated with the cation (Howlett et al. 2006).

During the cycling of Li‖Li symmetric cells with IL-based electrolytes, voltage instabilities were frequently noted (Basile et al. 2013; Best et al. 2010; Bhatt et al. 2010, 2012; Lane et al. 2010b). This was initially attributed to a "welding" mechanism for Li dendrites (Fig. 3.15) (Lane et al. 2010b), but later it was proposed that a Li electrode surface reorganization was instead occurring (Basile et al. 2013). This latter explanation is very plausible since a similar SEI layer disruption was proposed for the reduction in the polarization for Li plating/stripping with polymer electrolytes (Appetecchi and Passerini 2002). Unlike for a polymer electrolyte where the formation of a resistive interface is rather slow due to the lack of mobility of the electrolyte, for an IL-based electrolyte—especially one with reactive

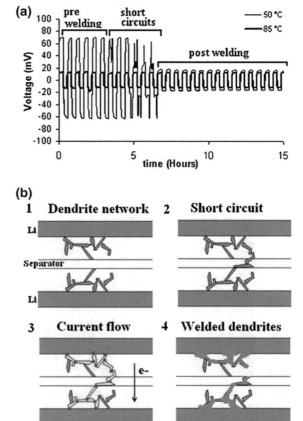

Fig. 3.15 a Time versus voltage curves for constant-current cycling of a Li‖Li symmetrical cell with a 0.4 M LiFSI-IL electrolyte with $MO_{14}FSI$ cycled at a current density of 0.2 mA cm^{-2} and **b** schematic illustration showing the steps in a possible dendrite welding process: *1* accumulated dendrites, *2* dendrites connect, causing a short circuit, *3* electrical current flows through granular conduction path creating hot spots at grain boundaries, and *4* short circuit removed due to a fuse effect with remaining dendrites unified into a single metallic structure with metallic connection to the base electrode. Reproduced with permission—Copyright 2010, The American Chemical Society (Lane et al. 2010b)

FSI^- anions or other electrolyte components—as the interphasial layer is disrupted, reactions with the exposed Li accelerate, resulting in an increase in the polarization. This buildup and disruption readily explains the chaotic polarization voltage variations during cycling.

3.2.5 Importance of Electrolyte Concentration

In recent years, use of highly concentrated electrolyte has proved to be a very efficient approach to suppress Li dendrite growth and enhance the CE of Li cycling. As the salt concentration increases, there is less free (uncoordinated) solvent present in the electrolyte. Thus, the SEI composition may be more reflective of the anion's reductive properties. However, in highly concentrated electrolytes there is not enough solvent available to fully coordinate the Li^+ cations. Thus, cations are coordinated to solvent molecules and anions, with the anions typically coordinated to multiple Li^+ cations, giving a network-like electrolyte solution structure. Just as coordinated solvent molecules are more susceptible to reduction, so too are coordinated anions.

Early work in this field was focused on the use of concentrated electrolyte to stabilize graphite electrodes used in Li-ion batteries. Recently, it has also been used in Li metal batteries with excellent results. Because graphite has a reduction potential very close to that of a Li metal anode, general characteristics of the concentrated electrolyte used with graphite are similar to those used for Li metal anodes. Therefore, it is very useful to understand the general properties of concentrated electrolytes when they are used with graphite.

McKinnon and Dahn demonstrated in 1985 that co-intercalation of PC into layered materials intercalated with Li can be reduced with concentrated electrolytes (McKinnon and Dahn 1985). McOwen et al. (2014) used highly concentrated electrolytes containing carbonate solvents with LiTFSI to determine the influence of eliminating bulk solvent (i.e., uncoordinated to a Li^+ cation) on electrolyte properties. The phase behavior of LiTFSI-EC mixtures indicates that two crystalline solvates form—LiTFSI-$(EC)_3$ and LiTFSI-$(EC)_1$. A Raman spectroscopic analysis of the EC solvent bands for the 3-1 and 2-1 LiTFSI- EC liquid electrolytes indicates that ~ 86 and 95 %, respectively, of the solvent is coordinated to the Li^+ cations. This extensive coordination results in significantly improved anodic oxidation and thermal stabilities as compared with more dilute (i.e., 1 M) electrolytes. Further, while dilute LiTFSI-EC electrolytes extensively corrode the Al current collector at high potential, the concentrated electrolytes do not. Although the ionic conductivity of concentrated LiTFSI-EC electrolytes is somewhat low relative to the current state-of-the-art electrolyte formulations used in commercial Li-ion batteries, using an EC-DEC mixed solvent instead of pure EC markedly improves the conductivity.

Recently, Yamada et al. (2010, 2013, 2014a, b) have done extensive work on use of various concentrated electrolytes to improve the stability of graphite anode under different conditions. They found that a superconcentrated ether electrolyte can lead to ultrafast Li^+ intercalation into graphite, even exceeding that in a currently used commercial electrolyte. This discovery is important for fast-charging Li-ion batteries far beyond present technologies (Yamada et al. 2013). They also investigated the behavior of electrochemical lithium intercalation into graphite in DMSO-based electrolytes, including lithium salt-concentrated solutions and binary solutions with DMC. Raman spectra of these solutions showed that the solvation number of DMSO molecules toward Li^+ cations (N_{DMSO}) decreased from 4.2 in a conventional solution to around 2 in the salt-concentrated solution and the binary solution. A comparison between the behavior of graphite and the N_{DMSO} values has clarified that the N_{DMSO} value of 3 is a criterion for determining whether the intercalation of Li^+ ions or solvated Li^+ ions occurs. A series of results suggests that the decrease in the solvation number of the relevant solvents toward Li^+ ion can suppress the co-intercalation of solvents into graphite (Yamada et al. 2010). They also found that a dilute electrolyte (e.g., 1 M) of sulfoxide, ether, and sulfone results in solvent co-intercalation and/or severe electrolyte decomposition at a graphite electrode, whereas their superconcentrated electrolyte (e.g., > 3 M) allows for highly reversible Li intercalation into graphite. A unique coordination structure in the super-concentrated solution and an anion-based inorganic SEI film on the cycled graphite electrode have been identified which would be the origin of the reversible graphite negative electrode reaction without EC (Yamada et al. 2014a).

Yamada and Tateyama (2014) also investigated the fundamental mechanism behind the enhanced stability of highly concentrated electrolyte toward graphite anode. They report enhanced reductive stability of a superconcentrated acetonitrile (AN) solution (>4 M). Applying it to a battery electrolyte, they demonstrate, for the first time, reversible Li intercalation into a graphite electrode in a reduction-vulnerable AN solvent. First-principle calculations combined with spectroscopic analyses reveal that the peculiar reductive stability arises from modified frontier orbital characters unique to such superconcentrated solutions, in which all solvents and anions coordinate to Li^+ cations to form a fluid polymeric network of anions and Li^+ cations. To identify the solution structure providing the unusual reductive stability with a surface film derived from bis(trifluoromethanesulfonyl)amide (TFSA or TFSI), Raman spectra were obtained for LiTFSI/AN solutions at various Li salt concentrations (Fig. 3.16a, b). In dilute solutions, a stable solvation structure around Li^+ is reported to be 3- or 4-fold coordination (Fig. 3.16c). Further increasing the Li salt concentration decreases free AN molecules and instead increases the Li^+-solvating AN molecules. At 4.2 M (super-high concentration), where unusual reductive stability was observed, there is only a peak for Li^+-solvating AN molecules, indicating that all the AN molecules coordinate to Li^+. A series of spectroscopic analyses show that the structure of the salt-superconcentrated solution is characterized by a fluid polymeric network of mutually interacting $TFSI^-$ anions and Li^+ cations in the presence of two AN molecules solvating each Li^+. This peculiar structural feature is unique to such a

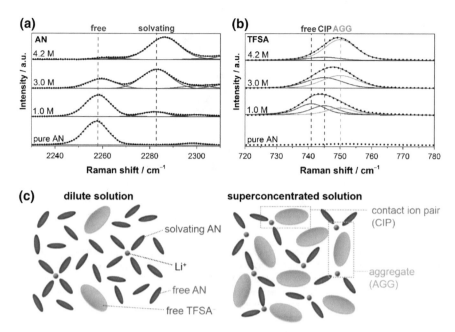

Fig. 3.16 Raman spectra of LiTFSI/AN solutions in **a** 2230–2310 cm^{-1} (C≡N stretching mode of AN molecules) and **b** 720–780 cm^{-1} (S–N stretching, C–S stretching, and CF$_3$ bending mode of TFSA). *Points* and *solid lines* denote experimental spectra and fitting curves, respectively. **c** Representative environment of Li$^+$ in a conventional dilute solution (i.e., ~1 M) and a salt-superconcentrated solution (i.e., 4.2 M). Reproduced with permission—Copyright 2014, American Chemical Society (Yamada et al. 2014b)

superconcentrated solution, and should be a clue to its modified film-forming ability that provides unusual reductive stability.

Yamada et al. further used quantum mechanical density functional theory molecular dynamics (DFT-MD) simulation to further elucidate the origin of the unusual reductive stability with a TFSI-derived surface film, DFT-MD was applied to the dilute (1-LiTFSI/43-AN, 0.4 M) and superconcentrated (10-LiTFSI/20-AN, 4.2 M) LiTFSI/AN systems. Panels a and b of Fig. 3.17 show snapshots and projected density of states (PDOS) of equilibrium trajectories of the dilute solution with a fully dissociated salt and contact ion pairs (CIPs), respectively. In the dilute system with a fully dissociated salt (Fig. 3.17a), the energy levels of Li$^+$-solvating or free AN molecules are lower than those of TFSI$^-$ anions at the lowest end of conduction bands; the LUMO is located at AN molecules. This indicates that AN molecules are predominantly reduced in a dilute LiTFSI/AN solution, which is consistent with the experimental result of continuous reductive decompositions of AN solvents in contact with Li metal, a strong reduction reagent. The situation is the same when a simple ion pair (i.e., CIP) exists in the dilute solution (Fig. 3.17b).

As for the superconcentrated solution (Fig. 3.17c), the snapshot at an equilibrium trajectory shows a polymeric network of TFSI$^-$ anions and Li$^+$ cations: each

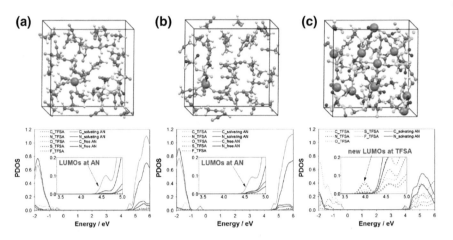

Fig. 3.17 Supercells used and PDOS obtained in quantum mechanical DFT-MD simulations on **a** and **b** dilute (1-LiTFSI/43-AN corresponding to 0.4 M) and **c** superconcentrated (10-LiTFSI/20-AN corresponding to 4.2 M) LiTFSI/AN solutions. The illustrated structures are the snapshots in equilibrium trajectories. For a dilute solution, both situations of LiTFSI salt [i.e., **a** full dissociation and **b** CIP] were considered. Atom color: Li *purple*; C *dark gray*; H *light gray*; O *red*; N *blue*; S *yellow*; F *green*. Li atoms are highlighted in size. *Insets* in the PDOS profiles are magnified figures of the lowest energy-level edge of the conduction band. Reproduced with permission—Copyright 2014, American Chemical Society (Yamada et al. 2014b)

TFSI$^-$ anion is interacting with multiple Li$^+$ cations to form ion pairs, which is consistent with the results of Raman analyses and totally different from those in dilute solutions. In the PDOS profile (see bottom of Fig. 3.17c), TFSI$^-$ anions provide new unoccupied states at the lowest energy in the conduction bands and the energy levels of TFSI$^-$ become lower than those of AN molecules; the LUMO is located at TFSI$^-$ anions in the superconcentrated solution. The new unoccupied states are unique to such a superconcentrated solution with characteristic structural features of mutually networking TFSI$^-$ anions and Li$^+$ cations with a small amount of solvents. The localized LUMOs at the TFSI$^-$ anions suggest that TFSI$^-$ anions, rather than AN solvents, are predominantly reduced to form a TFSI-derived surface film on a graphite electrode in the superconcentrated electrolyte. These quantum mechanical simulations with a realistic experimental salt concentration elucidate the peculiar frontier orbital characters in a salt-superconcentrated solution and account for the formation of a TFSI-derived surface film, which is the origin of the enhanced reductive stability that allows for reversible lithium intercalation into a graphite negative electrode.

With the successful use of highly concentrated electrolyte to enhance the stability of the electrolyte toward graphite electrode in the absence of EC, the idea of concentrated electrolyte has been used to enhance the stability of Li metal anodes recently. Tobishima et al. (1984) studied the influence of salt concentration for LiClO$_4$-PC/DME electrolytes up to a concentration of 3 M. The Li cycling CE plateaued at a high value of 92–93 % for 1 M and 2 M electrolytes, but decreased to 87 % for a 3 M concentration. Suo et al. (2013) reported that the use of LiTFSI-DME electrolytes up to a concentration of 7 M gave very promising Li-S cell performance due, in part, to the suppression of Li dendrite formation. These "solvent-in-salt" electrolytes exhibit a high Li$^+$-ion transference number (0.73), in which salt holds a dominant position in the Li$^+$-ion transport system. It remarkably enhances the cyclic and safety performance of next-generation high-energy rechargeable Li batteries via an effective suppression of Li dendrite growth and shape change in the Li metal anode. Moreover, when used in a Li-S battery, the advantage of this electrolyte is further demonstrated in that lithium polysulfide dissolution is inhibited, thus overcoming one of today's most challenging techno-logical hurdles, the "polysulfide shuttle phenomenon." Consequently, a CE nearing 100 % and long cycling stability are achieved.

Recently, Qian et al., found that replacing LiTFSI with LiFSI can greatly increase the stability of Li metal. The use of highly concentrated electrolytes composed of DME solvents and the LiFSI salt enables the high-rate cycling of a Li metal anode at high CE (up to 99.1 %) without dendrite growth. With 4 M LiFSI in DME as the electrolyte, a Li‖Li cell can be cycled at high rate (10 mA cm^{-2}) for more than 6000 cycles with no increase in the cell impedance, and a Li‖Cu cell can be cycled at 4 mA cm^{-2} for more than 1000 cycles with an average CE of 98.4 %, as shown in (Fig. 3.18). These excellent high-rate performances can be attributed to the increased solvent coordination and increased availability of Li$^+$ concentration in the electrolyte. The small amount of degradation that occurred during each cycle led to the slow buildup of a black interphasial layer, but this did not lead to a change in the polarization for the plating/stripping, suggesting that this layer did not impede the cycling to any significant extent. Further development of this electrolyte may lead to practical applications for Li metal anodes in rechargeable batteries. The fundamental mechanisms behind the high-rate ion exchange and stability of the electrolytes also shed light on the stability of other electrochemical systems.

3.2.6   Self-healing Electrostatic Shield Mechanism

As we discussed in the previous sections, various approaches have been used to suppress Li dendrite formation and growth. However, most of these approaches rely on the formation of a strong layer to mechanically block the Li dendrite growth. Recently, Ding et al. developed a novel mechanism that could influence Li depo-sition preference to obtain a dendrite-free Li anode, which is called the self-healing electrostatic shield (SHES) mechanism (Ding et al. 2013b). This mechanism does not rely on the mechanical strength of a protection layer. Instead, it relies on an electrostatic shield formed by electrolyte additives. This layer is positively charged and floating on the tip area of potential dendrites on the surface of deposited Li film,

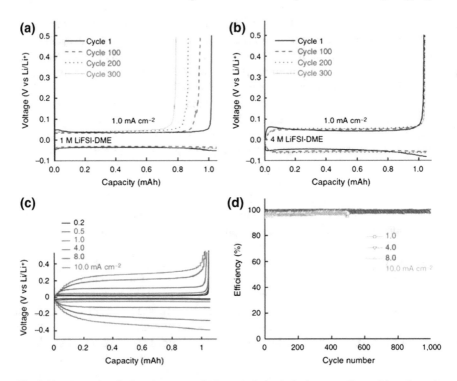

Fig. 3.18 Electrochemical performance of Li metal plating/stripping on a Cu working electrode: voltage profiles for the cell cycled in **a** 1 M LiFSI-DME and **b** 4 M LiFSI-DME, **c** polarization of the plating/stripping for the 4 M LiFSI-DME electrolyte with different current densities and **d** CE of Li deposition/striping in 4 M LiFSI-DME at different current densities: voltage profiles for the cell cycled in **a** 1 M LiFSI-DME and **b** 4 M LiFSI-DME, **c** polarization of the plating/stripping for the 4 M LiFSI-DME electrolyte with different current densities, and **d** CE of Li deposition/striping in 4 M LiFSI-DME at different current densities. Reproduced with permission—Copyright 2015, Macmillan Publishers Limited (Qian et al. 2015b)

but does not physically attach to the dendrite and Li film. During the Li deposition process, this electrostatic shield will repel the incoming Li$^+$ ions and force them to be deposited in the valley area instead of peak area of the Li film. Once applied electric voltage is turned off, this charged cloud will dissipate and does not form a permanent film on the Li surface.

Figure 3.19 illustrates the SHES mechanism to prevent Li dendrite formation. According to the Nernst equation, it is possible to find certain metal cations (M$^+$) that have an effective reduction potential lower than that of Li$^+$ if the M$^+$ has a standard redox potential close to those of Li$^+$ and a chemical activity much lower than that of Li$^+$ (Ding et al. 2013b). For example, cesium ion (Cs$^+$) at a low concentration of 0.01 M has an effective reduction potential of -3.144 V, which is lower than that of Li$^+$ at 1.0 M concentration (-3.040 V). As a result, in a mixed electrolyte where the additive Cs$^+$ concentration is much lower than the Li$^+$

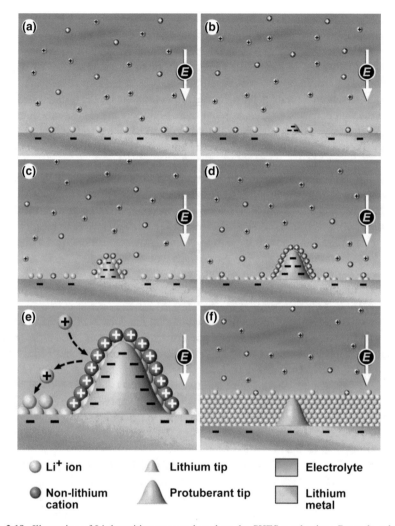

Fig. 3.19 Illustration of Li deposition process based on the SHES mechanism. Reproduced with permission. Copyright 2013, American Chemical Society (Ding et al. 2013b)

concentration, Cs^+ additives will not be deposited at the Li deposition potential and do not form thin layers of Li alloys at the electrode surface.

As shown in Fig. 3.19, at the initial stage of deposition, both Li^+ ions and the Cs^+ additive cations are adsorbed on the surface of the substrate (e.g., Li). When the applied voltage (V_a) is slightly lower than the reduction potential of a Li ion ($E_{Li/Li+}$) but higher than the reduction potential of the Cs^+ ($E_{Cs/Cs+}$), i.e. $E_{Li/Li+} > V_a > E_{Cs/Cs+}$, then Li deposits on the substrate surface (Fig. 3.19a). Due to the unevenness of the substrate surface or other fluctuations in the system, some Li protuberant tips will unavoidably be formed at the substrate surface (Fig. 3.19b).

Then a higher electron-charge density appears at the newly formed Li protuberance that attracts more Li ions from the electrolyte and deposits more on the tip than on other sites, thus forming the Li dendrite as would happen in the conventional electrolyte. However, in the SHES mechanism, more non-Li Cs^+ cations will also be adsorbed on the tip by electrostatic attraction force at the same time (Fig. 3.19c). When the adsorbed Li ions deposit on this tip continuously, the number of Li ions around the tip decreases significantly but more and more non-Li cations that do not electrochemically reduce at this potential will accumulate and surround the tip surface. Then a positively charged electrostatic shield is automatically formed by adsorbed non-Li cations on the Li tip (Fig. 3.19d). Due to the charge repelling and steric hindrance effect of the non-Li cation shield, it is more difficult for Li ions to adsorb and deposit on such a tip. Li ions have to deposit at other sites with lower non-Li cation adsorption ratios. As a result, the continuous Li growth on this tip (dendrite root) is stopped. It is reasonable to believe that there should be a lot of tips forming on the Li surface, but no tips would grow exceptionally as shown in Fig. 3.19e. Eventually, a smooth Li surface could be formed as illustrated in Fig. 3.19f; then this self-healing process will repeat itself. In fact, Fig. 3.19 is a simplified illustration of a Li deposition process with SHES additives. In practice, an SEI layer will always form on the surface of Li metal once it is in contact with electrolyte. Therefore, Cs^+ ions will accumulate on the outside of the SEI layer and quickly form an electrostatic shield and prevent amplification of Li dendrite growth.

The SHES additive approach is different from those of inorganic additives (including Mg^{2+}, Al^{3+}, Zn^{2+}, Ga^{3+}, In^{3+}, and Sn^{2+}) used in previous studies (Matsuda et al. 1991); (Matsuda 1993). These additive metal ions are consumed during each Li deposition and their effectiveness will soon fade with increasing cycles. However, in the proposed SHES mechanism, the additive cations will not be consumed and can last for long-term cycling. This prediction has been verified by chemical analysis of the deposited films.

Figure 3.20 shows SEM images of the deposited Li films on copper substrates with different Cs^+ concentrations. In the control electrolyte without the $CsPF_6$ additive, Li dendrite formation is clearly observed in the deposited Li film (see Fig. 3.20a). Even at very low Cs^+ concentrations (0.005 M), Li dendrite formation is significantly decreased (Fig. 3.20b). When the Cs^+ concentration is increased to 0.05 M, the dendrite formation is completely eliminated, resulting in a very distinct improvement (Fig. 3.20c). Further investigation indicated that even if a dendritic Li film was initially formed during Li deposition, it could be smoothened if further deposition was conducted in the electrolyte containing Cs^+ additive. In addition to Cs^+, Rb^+ also exhibits a lower effective reduction potential when its concentration is much lower than that of Li^+. Not surprisingly, $RbPF_6$ was also able to suppress Li dendrite growth although it is not as effective as $CsPF_6$. Unlike other cations, reported in the previous work, that will form part of the SEI layer (Matsuda et al. 1991) (Matsuda 1993), SHES additives (Cs^+, Rb^+, etc.) do not form part of the SEI layer as verified by several trace analysis techniques (Ding et al. 2013b).

Fig. 3.20 SEM images of the morphologies of Li films deposited in electrolyte of 1 M LiPF$_6$/PC with CsPF$_6$ concentrations of: **a** 0 M, **b** 0.005 M, **c** 0.05 M, at a current density of 0.1 mA cm^{-2}. Reproduced with permission. Copyright 2013, American Chemical Society (Ding et al. 2013b)

Finally, it is necessary to indicate that although the SHES mechanism has addressed the dendritic morphology problem encountered during Li deposition, the CE of Li deposition in the electrolytes used in the above work is still relatively low ($\sim$76 % in case of 1 M LiPF$_6$/PC). Further optimization of electrolyte solvent, salt, and additives have increased CE of Li deposition/stripping to more than 98 % and still retained dendrite-free morphology of the deposited Li films, which is required for long-term cycling operation of Li metal anodes.

3.3 Ex Situ Formed Surface Coating

Rather than using the electrolyte components to dictate the composition of in situ formed SEI layers on the Li, a wide variety of ex situ surface treatments and polymer coatings have been applied to the Li anode prior to cell assembly. Examples include: N$_2$ gas (generating Li$_3$N as noted above) (Wu et al. 2011), cyclopentadienyldicarbonyl iron (II) silanes (Fp-silanes) (Neuhold et al. 2014), Al$_2$O$_3$ (Kazyak et al. 2015), carbon thin films (amorphous, nanostructured, and diamond-like) (Zhang et al. 2014c; Bouchet 2014; Zheng et al. 2014; Arie and Lee 2011), Nafion (Song et al. 2015), polyvinylidene fluoride (PVDF)-Li$_2$CO$_3$ (Chung et al. 2003), PVDF-HFP (Jang et al. 2014; Lee et al. 2006; Osaka et al. 1999), PVDF-HFP-Al$_2$O$_3$ (Lee et al. 2015), polyacetylene (Belov et al. 2006), poly (vinylene carbonate-co-acrylonitrile) (P(VC-co-AN)) layer (Choi et al. 2013), plasma-polymerized 1,1-difluoroethene (Takehara et al. 1993), and crosslinked Kynar 2801-1,6-hexanediol diacrylate semi-IPN (interpenetrating polymer network) (Choi et al. 2003, 2004a, b). This last coating transformed the Li from a needle-like deposit to a nodular morphology (Fig. 3.21)—despite the use of the same electrolyte [i.e., 1 M LiClO$_4$-EC/PC (1/1)]—as the amount of the crosslinking agent (i.e., 1,6-hexanediol diacrylate) was increased.

Recently, Zheng et al. designed an interconnected hollow carbon nanospheres for stable Li metal anodes (Zheng et al. 2014). They developed a template synthesis

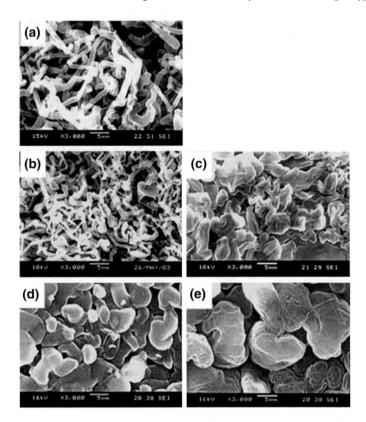

Fig. 3.21 SEM images of Li deposited from a 1 M LiClO₄-EC/PC (1/1) electrolyte **a** without a surface coating, **b** with a protective coating of only Kynar 2801 and with semi-IPN surface coatings with Kynar 2801/1,6-hexanediol diacrylate ratios: **c** (7/3), **d** (5/5) and **e** (3/7). Reproduced with permission—Copyright 2004, Elsevier (Choi et al. 2004b)

method for fabricating the hollow carbon nanopheres, using vertical deposition of polystyrene nanoparticles (Fig. 3.22a). A colloidal multilayer opal structure is formed on Cu foil by slowly evaporating a 4 % aqueous solution of carboxylated polystyrene particles. The highly monodisperse polystyrene nanoparticles form a close-packed thin film with long range order (Fig. 3.22b). The polystyrene nanoparticles are coated with a thin film of amorphous carbon using flash-evaporation of carbon fibers (Fig. 3.22c). The samples are then heated in a tube furnace to 400 °C under an inert atmosphere, forming hollow carbon nanopheres on the Cu substrate (Fig. 3.22e). TEM characterization shows that the carbon wall has a thickness of ∼ 20 nm (Fig. 3.22f). The hemispherical carbon nanospheres are interconnected to form a thin film, which can be peeled off the Cu surface easily. Mechanical flexibility is important in accommodating the volumetric

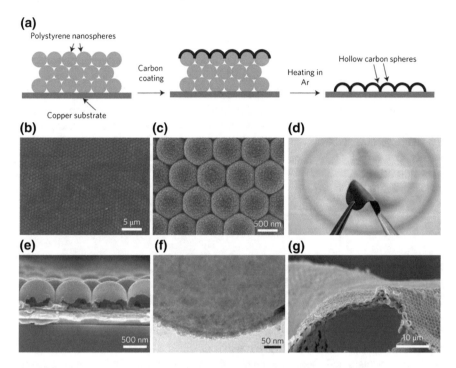

Fig. 3.22 Fabrication of hollow carbon nanosphere-coated electrode. **a** Fabrication process for the hollow carbon nanosphere-modified Cu electrode. *Left to right* Polystyrene nanoparticles are first deposited onto the Cu substrate; a thin film of amorphous carbon is coated on *top* of the polystyrene array using flash-evaporation of carbon cord; thermal decomposition of the polystyrene template results in the formation of interconnected hollow carbon nanospheres. **b, c,** SEM images of the carbon-coated polystyrene nanoparticle array at low (**b**) and high (**c**) magnifications. The slight morphology change of the carbon nanospheres to a hexagonal shape could be due to the elevated temperature during the carbon-coating process. **d** Digital camera image of the as-fabricated hollow carbon nanosphere thin film after removal of the polystyrene template. **e** Cross-sectional SEM image of the hollow carbon nanospheres. **f** TEM image of the hollow carbon nanospheres, with wall thickness of ∼20 nm. **g** SEM image of the hollow carbon nanosphere thin-film peeled off the Cu substrate. *Red dashed line* trace of the curvature of bending. Reproduced with permission—Copyright 2014, Nature Publishing Group (Zheng et al. 2014)

change of Li deposition and dissipating the stress exerted on the Li protection layer during cycling. A digital camera image (Fig. 3.22d) and SEM image (Fig. 3.22g) show that the carbon nanosphere thin film can achieve a bending radius of ∼20 m. They demonstrate that coating the lithium metal anode with such a monolayer of interconnected amorphous hollow carbon nanospheres helps isolate the Li metal depositions and facilitates the formation of a stable SEI. The CE of Li cycling with such an ex situ formed surface layer is ∼99 % for more than 150 cycles. This result indicates that nanoscale interfacial engineering could be a promising strategy to tackle the intrinsic problems of lithium metal anodes.

3.4 Mechanical Blocking and Solid Electrolytes

Commercial separators for Li-ion batteries tend to be thin (<30 µm), microporous polyolefin membranes made from polyethylene (PE), polypropylene, or laminates of PE/polypropylene (Arora and Zhang 2004). Such separators are frequently also used when testing cells with Li anodes. It has been proposed, however, that dendrite formation can be inhibited by incorporating a solid separator/electrolyte, especially one with an acceptable ionic conductivity and a high shear modulus of approximately twice that of Li (i.e., 9 GPa) (Monroe and Newman 2005; Ferrese and Newman 2014).

3.4.1 Solid Polymer Electrolytes

Early work with solid polymer electrolytes (SPEs) and Li metal anodes focused on Li salt-PEO salt and related electrolytes (Gauthier et al. 1985b; Xue et al. 2015). The relatively low conductivity of such electrolytes and their tendency to crystallize generally required the cells to be plasticized via the inclusion of a liquid electrolyte or instead operated at elevated temperature (e.g., 75–90 °C) (Appetecchi et al. 1998). Using a polyether copolymer with an added 1–3 M $LiClO_4$-EC/PC (1/1) electrolyte, Matsui and Takeyama used a visual study of the Li-plasticized polymer electrolyte interface to examine the effects of the electrolyte composition on the Li deposition morphology (Matsui and Takeyama 1995). The solid electrolytes that most effectively suppressed Li dendrites were those with a lower liquid electrolyte content (≤ 70 %) and that were stiff and elastic (e.g., with a molecular weight > 16,000 before crosslinking).

Unplasticized polyether electrolytes are generally found to exhibit a low Li/SPE interfacial resistance that remains relatively stable over time, even upon storage at elevated temperature (Appetecchi et al. 1998; Appetecchi and Passerini 2002; Fauteux 1993). Importantly, upon discharging the Li anode, the passivation layer formed at this interface is disrupted as indicated by the Li plating/stripping overvoltage dropping to a low value (Appetecchi and Passerini 2002). The ionic resistance of this layer, however, increased substantially with decreasing temperature (≤ 65 °C) due to the crystallization of some of the PEO ($T_m \sim 65$ °C) (Appetecchi and Passerini 2002). This prevents self-discharge and other parasitic reactions during prolonged storage at ambient temperature, while the rapid disruption of the layer after a few initial cycles at elevated temperature enables the cells to perform well even after prolonged storage periods. The cell assembly procedure (including hot lamination) used for the Li/SPE interface was found to be critical for determining the stability of the interface over time (Appetecchi et al. 2000). In situ attenuated total reflectance FTIR measurements have indicated that electrodeposited Li reacts with the PEO in a $LiClO_4$-PEO electrolyte to form alkoxides (ROLi) (Ichino et al. 1991), similar to the reaction of Li with glymes (Aurbach and Granot 1997) and THF:

$$—CH_2OCH_2— + Li \rightarrow —CH_2OLi + other\ products$$

After an initial fast reaction period, further reaction with polyethers is relatively slow (Lisowska-Oleksiak 1999). For SPEs containing LiBETI, the interfacial resistance was found to proportionally depend upon the salt concentration (at open circuit) (Appetecchi and Passerini 2002). This indicates that the salt may have a key role in the electrode passivation (either by reacting directly with the Li metal or influencing the reaction of the PEO with the Li). Evaluations of Li surfaces exposed to polyether network polymers (NPs) containing either LiTFSI or $LiBF_4$ indicated that the Li-polymer electrolyte interfacial resistance decreased and then became relatively stable for the electrolyte with LiTFSI, but the resistance grew continuously and became very large for the $LiBF_4$-containing electrolyte (Ismail et al. 2001). XPS indicated that the SEI layers formed in the LiTFSI electrolyte were relatively thin with $TFSI^-$ components present at the solution interface (either due to adsorbed or degraded anions), while the layers formed with the $LiBF_4$ electrolyte were thicker and contained much more LiF (Fig. 3.23) (Ismail et al. 2001). These results are analogous to what is obtained for liquid ether electrolytes in which the high stability of the ether solvent results in SEI compositions that are governed by the anion reactions.

Contrary to what was initially believed, Li dendrites do form in polymer electrolytes (Fig. 3.24a–c) (Brissot et al. 1998, 1999c, d; Dollé et al. 2002; Liu et al. 2010a, b, 2011b; Wang et al. 2012). To improve the ionic conductivity and restrict

Fig. 3.23 Schematic illustration of the surface film formed on Li foil after contact with **a** LiTFSI-NP and **b** $LiBF_4$-NP polymer electrolytes. Reproduced with permission—Copyright 2001, Elsevier (Ismail et al. 2001)

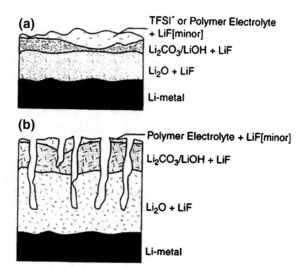

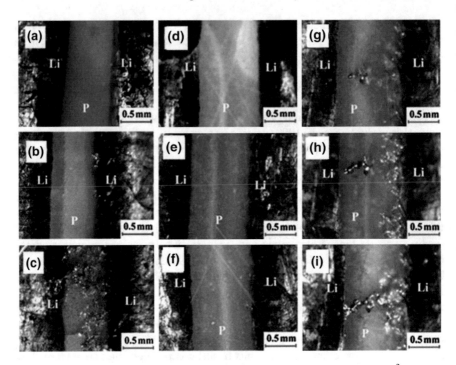

Fig. 3.24 Optical images of Li dendrite growth at a current density of 0.5 mA cm^{-2} at 60 °C. *First column* Li|LiTFSI-P(EO)$_{18}$|Li cell at times of **a** 0, **b** 15, and **c** 20 h *second* and *third column* Li|P(EO)$_{18}$-LiTFSI-1.44PI$_{13}$TFSI|Li cell at times of **d** 0, **e** 30, **f** 35, **g** 45, **h** 65, and **i** 75 h. Reproduced with permission—Copyright 2010, The Electrochemical Society (Liu et al. 2010b)

Li dendrite growth in polymer electrolytes, the polymers have been crosslinked (Khurana et al. 2014; Ueno et al. 2011), inorganic particle fillers have been added (Appetecchi et al. 1998, 1999; Zhang et al. 2002; Liu et al. 2010a; Kim et al. 2013c) and ILs have been incorporated (Shin et al. 2003, 2005a, b, 2006; Cheng et al. 2007; Choi et al. 2007b, 2011; Kim et al. 2007a, 2010a Zhu et al. 2008; Appetecchi et al. 2010, 2011; Balducci et al. 2011; Yun et al. 2011; Kim et al. 2012; Wang et al. 2012; Wetjen et al. 2013; Swiderska-Mocek and Naparstek 2014; de Vries et al. 2015). Li cycling CE values >95 % were reported for cells with SPEs with and without fillers, and the cells could be cycled for many hundreds of cycles at elevated temperature (e.g., 90 °C) (Appetecchi et al. 1998, 1999, 2000; Appetecchi and Passerini 2002; Gauthier et al. 1985a). The addition of ILs to SPEs enabled the operating temperature to be lowered to 30–40 °C and reduced the tendency for Li dendrites to form and penetrate the SPE (Fig. 3.24d–i) (Shin et al. 2003, 2005b; Liu et al. 2010b). The inclusion of both an inorganic filler and IL did enable repeated Li plating/stripping cycles without dendrite formation (Liu et al. 2011b).

To further increase the shear modulus of the SPEs to prevent dendrite propagation, block copolymers have been studied in which a polystyrene component creates a rigid structure, while a PEO component dissolves a Li salt to create an ionic transport percolated pathway. These were first reported by Niitani et al. for polystyrene (PS)-PEO-PS (ABA) block copolymers in which the PEO segments were grafted to a PS backbone forming a percolated network structure (Fig. 3.25) (Niitani et al. 2005a, b, 2009). This was later pursued by Balsara et al. and Epps et al. using PS-PEO (AB) diblock copolymers (Singh et al. 2007; Young et al. 2008, 2011, 2014; Wanakule et al. 2009; Teran and Balsara 2011; Young and Epps 2012; Stone et al. 2012; Teran et al. 2012; Hallinan et al. 2013; Gurevitch et al. 2013; Schauser et al. 2014; Devaux et al. 2015a, b). Such polymers have been dubbed SEO (i.e., styrene-ethoxy) SPEs. Since the conductive phase is only a fraction of the material, SEO electrolytes have a lower conductivity than comparable LiX-PEO electrolytes and, in addition, the PEO segments may crystallize (as PEO or as high melting LiX-PEO crystalline phases), (Young et al. 2008) thus requiring that cells with SEO SPEs be operated at elevated temperature ($\geq 80\ ^{\circ}$C), and slow Li plating rates are required (Devaux et al. 2015b). It was reported that full Li|SEO|LiFePO$_4$ cells had an efficiency of >99 % for slow discharge rates (for the cathode) (Hallinan et al. 2013; Devaux et al. 2015a, b), but an examination of cycling voltammetries (CVs) for the Li plating/stripping indicates that significant current was consumed for reduction reactions well before 0 V versus Li/Li$^+$ during the cathodic scans, with only a small portion of this recovered during the subsequent anodic scans (Devaux et al. 2015b). In addition, the charge associated with the Li plating (below 0 V vs. Li/Li$^+$) was much higher than that for stripping suggesting that (at least for the CV measurements) the CE for the Li cycling may be quite low (Devaux et al. 2015b). Although such SPEs do restrict dendrite formation, they do not prevent dendrites which ultimately do penetrate the electrolyte, resulting in the short circuit (low impedance) failure of the cells (Fig. 3.26) (Stone et al. 2012; Hallinan et al. 2013; Gurevitch et al. 2013; Schauser et al. 2014). It was noted that this failure mode was common for Li|SEO|Li symmetric cells, whereas capacity fade due to delamination of the Li from the SEO SPE instead occurred for

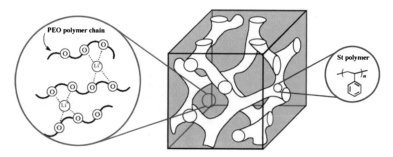

Fig. 3.25 Schematic illustration of an SEO block copolymer. Reproduced with permission—Copyright 2005, Elsevier (Niitani et al. 2005a)

Fig. 3.26 **a** Typical voltage versus charge passed prior to a short circuit (C_d) data showing the short circuit of the cell at 187 C cm^{-2} for a Li|SEO|Li cell cycled at 90 ° C with a current density of 0.17 mA cm^{-2}. **b** Cycling data showing C_d as a function of PEO molecular weight. **c** C_d as a function of storage modulus, G' (modulus determined for polymer without Li salt at 10 rad s^{-1} and 90 °C). Symbols correspond to SEO electrolytes (*Filled diamond*) and PEO electrolytes (*Filled square*). Reproduced with permission—Copyright 2012, The Electrochemical Society (Stone et al. 2012)

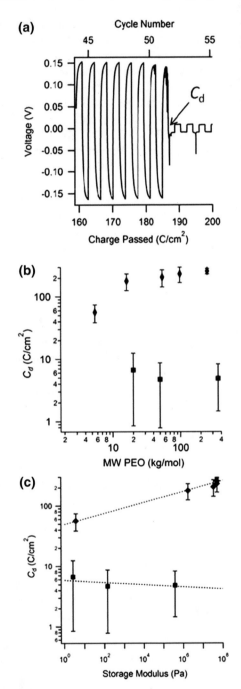

Li|SEO|LiFePO$_4$ cells (Devaux et al. 2015a). One possible explanation for this is that when the latter cells are cycled, the plating of the Li will result in an expansion of the cell (since the volume increase for the plated Li is significantly larger than the volume decrease for the cathode delithiation). Discharge of the cell then results in a contraction. The repetition of this expansion/contraction during cycling likely results in the delamination, since the SEO SPE is rigid. In contrast, when Li is simply shuttled from one Li electrode to another (i.e., symmetric cell), the cell volume should remain approximately constant during cycling and Li dendrites will extend into the electrolyte from both electrodes through defects in the rigid SPE. Since high modulus (i.e., rigid) block copolymer electrolytes have been unsuccessful at fully preventing dendrites from penetrating across the electrolyte to form short circuits, solid electrolytes with an even higher modulus, such as inorganic (crystalline or glassy) electrolytes, have been developed for batteries with Li anodes.

3.4.2 Solid Inorganic Electrolytes

Inorganic SSEs composed of inorganic compounds are nonflammable and generally more electrochemically stable (Jung et al. 2015; Takada 2013; Fergus 2010; Ren et al. 2015b). Therefore, they have been regarded as the ideal materials to protect Li metal anodes against dendrite penetration because of their unity Li$^+$ transfer number ($t_{Li}+$) and high mechanical strength, which often far exceeds that of Li itself. Both thin films and bulk forms of solid-state Li-ion conductors have been developed to effectively block Li dendrite growth (Christensen et al. 2012; Shao et al. 2012b). To date, the most widely used inorganic thin-film ion conductor is nitrogen-doped Li-ion phosphate film (LiPON) developed by Bates and Dudney et al. in the early 1990s (Bates et al. 1993). LiPON exhibits a conductivity of 2×10^{-6} S cm^{-1} at 25 °C and excellent long-term stability in contact with Li metal. Bates and Dudney also first reported the application of LiPON as a Li-ion-conducting electrolyte and Li metal protection layer in a thin-film battery (Bates 1994). Later on, Herbert et al. (2011) reported that the shear modulus of LiPON is approximately 77 GPa, 7.3 times higher than that of Li, which far exceeds the basic requirement of mechanical strength for electrolyte to suppress the Li dendrites, about twice that of Li. This result is also independent of the substrate type, film thickness, and annealing; therefore, LiPON is expected to be fully capable of mechanically suppressing dendrite formation at the Li/LiPON interface in thin-film batteries.

The typical structure of thin-film, solid-state batteries using LiPON as ion conductor is Li|LiPON|LiCoO$_2$ (Bates et al. 1993). The long cycle and shelf lives of these batteries result from the properties of the glassy LiPON electrolyte, which is stable in contact with metallic Li at potentials from zero to nearly 5.5 V and has acceptable conductance in the thin-film form. These batteries have demonstrated a very long shelf life. For example, Dudney (2005) demonstrated that test cells have maintained a charged state for more than a year with negligible loss of capacity

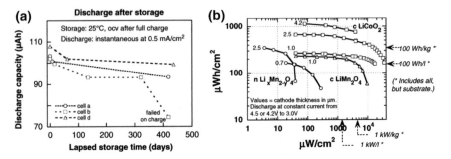

Fig. 3.27 a Capacities recorded for three Li∥LiCoO₂ thin-film batteries when rapidly discharged after prolonged storage in the fully charged state. **b** Power and energy density determined from constant-current discharge measurements for thin-film batteries with a Li anode and the indicated thin-film cathode. The batteries were discharged over 4.2–3.0, 4.5–3.0, and 4.5–2.5 V for the cLiCoO₂, cLiMn₂O₄, and nLiₓMn₂₋ᵧO₄ cathodes, respectively (*Filled square*). Reproduced with permission—Copyright 2005, Elsevier (Dudney 2005)

(see Fig. 3.27a. This level of reliability is not possible for other types of rechargeable batteries. These batteries also exhibit very high specific energies (Wh/kg) and specific power, as shown in Fig. 3.27b when only the active materials are considered. The effect of substrate and package material on the specific energy and energy density of the batteries will be discussed later. The cycle life of these batteries can exceed more than 40,000 full depth charge–discharge cycles, which far exceeds the cycle lives for other types of batteries. There is no liquid electrolyte, no polymer, or any other organic material present in thin-film, solid-state batteries; therefore, side reactions between electrode and electrolyte are minimized.

Many different bulk-form ceramic glass (∼50–200 μm thick) Li-ion conductors have also been developed and can effectively suppress Li dendrite growth. One example of these glass electrolytes is LiSICON-type $Li_{1+x}Al_xTi_{2-x}(PO_4)_3$ (LATP) developed by Fu (1997a, b). Recently, Wang et al. reported a Li∥LiMn₂O₄ battery operated in aqueous electrolyte using a Li metal anode protected by LATP glass. The battery demonstrates excellent stability and good electrochemical performance (Wang et al. 2013c). LATP glass prepared by Ohara glass and other companies has been widely used by many research groups worldwide to protect Li metal glass and has been applied in Li-air and Li-S batteries, as well as other energy storage and conversion systems. LATP glass is stable in weak acid and alkaline electrolyte. One of the disadvantages of LATP glass is that it is not stable when in contact with Li metal. Visco et al. first solved this problem by introducing an interfacial layer (a solid layer such as Cu₃N, LiPON, or nonaqueous electrolyte) between the Li metal and the Ohara glass, thus forming a protected Li electrode (PLE) (Visco et al. 2004a, 2009). Figure 3.28 shows the schematic of a PLE proposed by Visco et al. (2009), where the Li electrode was protected by an interfacial layer and a Li metal phosphate glass. Figure 3.26b shows a Li-air battery with a double-sided PLE. Li-air batteries using this PLE can operate in both aqueous and nonaqueous electrolytes (Zhang et al. 2010b; Yoo and Zhou 2011; Li et al. 2012a, b). However,

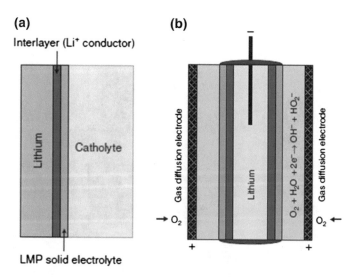

Fig. 3.28 **a** Use of an interlayer and a water-stable solid electrolyte to protect Li (*LMP* Li metal phosphate). **b** Schematic of a Li-air battery based on a PLE technology. Reproduced with permission—Copyright 2009, Elsevier (Visco et al. 2009)

large-scale application of these inorganic solid-state ceramic Li ion conductors is still hindered by their high cost, poor mechanical stability and limited ionic conductivity. Further development of new, flexible, inorganic, solid-state Li-ion conductors with good mechanical strength and stability, high Li ionic conductivity, excellent compatibility with Li metal, and wide electrochemical windows is still under investigation for their application in rechargeable Li metal batteries.

Another extensively studied type of solid-state inorganic electrolytes is garnet-like oxide glasses ($Li_6ALa_2Ta_2O_{12}$ [A = Sr, Ba]) developed by Thangadurai and Weppner (2005; Thangadurai et al. 2003; Murugan et al. 2007). Total conductivity values as high as 1.6 mS cm^{-1} have been achieved (Du et al. 2015). Also, they have been shown to be stable with Li metal (Cheng et al. 2015a). Although Li dendrites have been observed in garnet oxides, this is likely to be the results of defects (e.g., grain boundaries) in the samples rather than intrinsic mechanical problems of garnet oxides (Ren et al. 2015a; Ishiguro et al. 2013).

Kamaya et al. (2011) reported a superionic conductor with a composition of $Li_{10}GeP_2S_{12}$ which exhibited an extremely high Li$^+$ ionic conductivity of 12 mS cm^{-1}. This is much higher than the traditional solid-state electrolytes such as oxide perovskite, $La_{0.5}Li_{1.5}TiO_3$, thio-LISICON, or $Li_{3.25}Ge_{0.25}P_{0.75}S_4$, with ionic conductivities of the order of 10^{-3} S cm^{-1}. They found highly anisotropic conduction of Li$^+$ ions along one crystal direction and derived low activation energy for ionic conduction in this new material (see Fig. 3.29). Unlike traditional Li_3N or $LiTi_{1.7}Al_{0.3}(PO_4)_3$, the electrochemical window of $Li_{10}GeP_2S_{12}$ is wide (vs. Li/Li$^+$); thus, it is suitable for many different kinds of anode or cathode materials. When used in an In‖LiCoO$_2$ cell, excellent reversibility has been demonstrated with

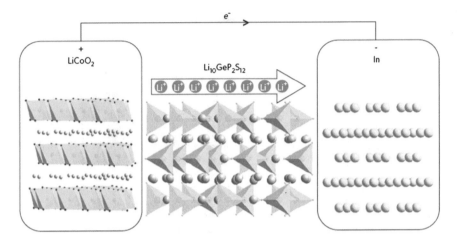

Fig. 3.29 All-solid-state electrolyte reported by Kamaya et al. Li$^+$ ions transport quickly through partially occupied LiS$_4$ tetrahedra in the Li$_{10}$GeP$_2$S$_{12}$ lattice and interstitial positions that are connected by a common edge. Reproduced with permission—Copyright 2011, Macmillan Publishers Limited (Kamaya et al. 2011)

operation voltage at 3.3 V. One disadvantage of Li$_{10}$GeP$_2$S$_{12}$, as well as most of sulfur-based Li ionic conductors, is that they are highly sensitive to moisture and unstable when in contact with Li metal. Therefore, they cannot be used in direct contact with Li anodes and their processing requires a very low humidity environment. Since they are even more sensitive to moisture than Li metal itself, so they often need to be processed in an Ar-filled glove box rather than in a dry room, which is suitable to handle Li metal.

Although solid, inorganic electrolytes have many advantages, including suppressing dendrite, stability with Li anodes, reducing battery capacity losses from cycling (i.e., a longer cycle life), usability at elevated temperature, improved safety, and high reliability, while other promoted features include the simplicity of cell design, an absence of leakage, a better resistance to shocks/vibrations and aggressive environments, and superior electrochemical, mechanical, and thermal stability; (Fergus 2010; Cao et al. 2014; Wang et al. 2015c; Knauth 2009), seldom does a solid-state electrolyte exhibits all these advantages at the same time. Therefore, the practical applications of the solid-state electrolyte are often limited to thin films and associated with other limiting factors.

In thin-film solid-state batteries, the glass electrolyte is used only to separate the anode and cathode. The energy density of solid-state batteries is still very limited because the thickness of electrode is limited by its own ionic and electronic conductivity. Through improvements in the ionic conductivity and stability of SSEs, these electrolytes also may be used as an ionic conductor inside thick electrode layers. Some SSEs have been used with limited success to prepare solid-state batteries (Nishio et al. 2009). In addition to good ionic conductivity of the electrolyte itself, several other conditions need to be optimized for practical applications of high-energy density miniature energy-storage devices. These additional conditions include minimizing the contact resistance between the solid electrolyte and the electrode particles, reducing the amount of SSE used in the batteries, and improving the compatibility of the electrolyte and electrode materials. Most of the existing SSEs cannot satisfy all of these conditions at the same time as analyzed below:

3. *Electrolyte/Electrode Stability.* Most highly conductive SSEs have limited electrochemical stability windows. That is, they are either unstable against a highly oxidative cathode or against highly reductive anodes, as in the case of $La_{0.56}Li_{0.33}TiO_3$, which involves the reduction of $La_{0.56}Li_{0.33}TiO_3$, the reduction of Ti^{4+} ions accompanied by the insertion of Li^+ ions. On the other hand, a sulfide-based electrolyte (such as $70Li_2S \cdot 29P_2S_5 \cdot P_2S_3$) and $Li_{1.5}Al_{0.5}Ge_{1.5}(PO_4)_3$ is not stable with a Li anode.

4. *Large Interface Impedance between Electrolyte and Electrode.* Unlike in the case when liquid or polymer electrolytes are used, there is no intimate contact between the electrode and the solid-state electrolyte. For example, $LiCoO_2$ cathode particles have significantly different particle sizes and physical/chemical properties when compared to sulfide-based solid-state electrolytes so ions and electrons are difficult to transfer between the electrode and the electrolyte. Furthermore, this interface impedance may increase with increasing cycle numbers because of mechanical instability (i.e., expansion/shrinkage) of the electrode that occurs during charge/discharge process.

5. *Low Capacity and Specific Energy of Full Device.* Nagao et al. (2009) and other groups have investigated batteries that are entirely solid state using a sulfide-based electrolyte. However, the cathodes ($Cu_xMo_6S_{8-y}$) used by Nagao et al. in their batteries have limited capacities (<100 mAh/g). They also used a large amount of electrolyte ($\sim$60–80 %) in the cathode to improve the conductivity of the electrode. Therefore, the capacities of the complete batteries are much less than those of conventional Li-ion batteries. It is not surprising that only the specific capacity of the active material in the cathode has been reported in the most of publications on bulk Li-ion batteries that are using entirely inorganic solid-state materials.

Another challenges for solid-state electrolytes is the difficulties associated with the high-temperature processing required to remove porosity for the production of very thin, rigid electrolyte sheets that are defect-free (Knauth 2009; Baek et al. 2014; Nemori et al. 2015; Imanishi et al. 2008; Cheng et al. 2014). In general, solid inorganic conductors may be prepared with a high ionic conductivity or a high electrochemical stability, but not both (Thangadurai and Weppner 2006). Thus, solid inorganic electrolytes are often coated with an "interlayer" of an ion-conducting solid such as a polymer electrolyte, Li_3N, or $LiBH_4/LiI$ to prevent direct contact between the solid inorganic electrolyte and Li, (Sahu et al. 2014; Wang et al. 2013a; Zhang et al. 2008, 2009, 2010c, 2011) or alternative anodes such as a Li–In alloy are used instead (Sahu et al. 2014; Zhang et al. 2008, 2009, 2010c, 2011; Wang et al. 2013a) or alternative anodes such as a Li-In alloy are instead used (Minami et al. 2006; Tatsumisago et al. 2006; Kamaya et al. 2011). Recently, however, Liu et al. reported that a Li_3PS_4 electrolyte had a high conductivity and excellent stability when cycled with a Li anode, but only a few plating/stripping cycles were shown and the cell polarization increased with each cycle (Liu et al. 2013). In addition, despite the high conductivities that have now been achieved ($>10^{-3}$ S cm^{-1}) and a Li$^+$ transference number of unity, the high-rate capability of cells with solid inorganic electrolytes is often inferior to those with liquid electrolytes due to a high contact resistance between the solid electrolyte grains, as well as with the solid electrode materials (Ohta et al. 2006; Sakuda et al. 2010).

As a result of these problems, only a few publications are available about the Li cycling CE for solid inorganic electrolytes, as these publications often instead focus on the electrolyte–cathode interface (Ohta et al. 2012, 2013), but what has been published indicates that the CE values are <100 %, often well below this, and that dead Li may form due to side reactions (Motoyama et al. 2015; Sagane et al. 2013a; Neudecker et al. 2000; Yamamoto et al. 2012). In addition, large volume changes occur during Li deposition/stripping and when Li is plated onto a current collector in contact with a rigid solid electrolyte, the Li may deform either the current collector or the electrolyte (Motoyama et al. 2015; Neudecker et al. 2000; Sagane et al. 2013b). During deposition at higher current densities (>1 mA cm^{-2}), Li plating leads to Li growth in the pores and along the grain boundaries of the solid electrolyte (Nagao et al. 2013a). This results in crack propagation which facilitates (Bates et al. 1993) Li dendrite growth through the solid electrolyte and the eventual short circuit of the cell (Nagao et al. 2013a). Several reports have confirmed that short-circuiting occurs through grain boundaries and interconnected pores (Sudo et al. 2014; Suzuki et al. 2015; Ren et al. 2015a). Inherent defects within the solid electrolytes from processing may thus limit the ability of these electrolytes to prevent short-circuiting. Even if such initial defects can be eliminated, the mechanical stresses induced during cycling from Li growth and volume changes are likely to generate and grow new defects. Therefore, extensive research still needs to be done before broad commercial application of inorganic solid-state electrolytes.

3.5 Effect of Substrates

The substrate used as a working electrode for the electrochemical reduction of Li$^+$ cations is important (Fig. 3.30) because in many cases the Li$^+$ cations react with the material to form alloys rather than forming Li metal on the substrate's surface (Dey 1971; Huggins 1988, 1999a, b, 1989a). For non-Li metal working electrodes, the surface layers formed when the metal surface atoms react with contaminants or undergo conversion reactions with Li_2O may also complicate the electrodeposition chemistry. Surface roughness is also a key consideration for SEI formation and the way Li deposits on the substrate.

Fig. 3.30 Cyclic voltammograms (CVs) (first cycle) of Ni and Al electrodes with a 1 M $LiClO_4$-PC electrolyte (5 mV s^{-1}). Reproduced with permission —Copyright 1981, Elsevier (Frazer 1981)

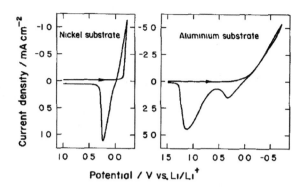

3.5.1 Alloys

Alloys are mixtures of metals (or a metal and another element) with metallic bonding. They may consist of solid solutions (single phase) or mixtures of phases. For many Li-based alloys, intermetallic compounds form with specific stoichiometries and unique crystal structures. Li can thus be inserted into other materials and, by avoiding dendritic growth in which the reactive Li metal is directly exposed to the electrolyte, much higher cycling efficiencies can be obtained (Sazhin et al. 1994). Brief descriptions are provided below for the known phase behavior for Li with carbon (graphite) and metals that are commonly used as electrodes.

3.5.1.1 Li-C (Graphite)

Commercial Li-ion batteries are based upon anodes in which Li^+ cations intercalate into carbon (typically graphite) to form Li-C alloys (Huggins 2009; Ogumi and Wang 2009; Okamoto 1989). The phase diagram for Li-C(graphite) indicates that multiple phases form: $C_{72}Li$, $C_{36}Li$, $C_{18}Li$, $C_{12}Li$, C_6Li, and αCLi (Okamoto 1989). The progressive intercalation of Li^+ cations into graphite during the reduction reaction is referred to as staging, and a key foundation of Li-ion batteries is the reversibility of this reaction up to the C_6Li phase (thus stipulating the maximum reversible capacity achievable for a graphite electrode).

3.5.1.2 Li–Cu

Cu, the most commonly used current collector for graphite-based anodes for Li-ion batteries, is frequently used as a working electrode for the plating/stripping of Li metal. Li has a fairly high solubility in Cu when melt is processed at high temperature, mechanically alloyed (i.e., ball milled) or when electrodeposited from molten salts (Baretzky et al. 1995; Gąsior et al. 2009; Pastorello 1930; Camurri et al. 2003; Lambri et al. 1996, 1999, 2000, 2005; Pérez-Landazábal et al. 2002; Peñaloza et al. 1995). Variable-composition Li-Cu solid solution alloys can thus be formed at elevated temperature up to a composition of approximately 16–18 mol% Li (Pelton 1986c; Gąsior et al. 2009). At lower temperatures, however, there are no indications that Li–Cu solid solutions readily form. Li can thus be plated/stripped on Cu with very high efficiency, although it is possible that some limited reaction occurs between the Li and Cu at the interface, especially during the first Li plating step.

3.5.1.3 Li–Al

Li reacts readily with Al (Armstrong et al. 1989; Besenhard 1978; Besenhard et al. 1985, 1990; Carpio and King 1981; Fischer and Vissers 1983; Frazer 1981; Fung and Lai 1989; Garreau et al. 1983; Geronov et al. 1984a, b, c; Hamon et al. 2001; Huggins 1999b; Jow and Liang 1982; Morales et al. 2010; Myung et al. 2010; Okamoto 2012b; Park et al. 2002; Rao et al. 1977; Suresh et al. 2002; Zhou et al. 2010; Zlatilova et al. 1988; Myung and Yashiro 2014). Figure 3.31 shows that—at 423 °C—Li begins to react above 0.35 V (vs. Li/Li$^+$) to form a solid solution with Al (α phase) up to a concentration of close to 10 atom% (or at.%). This is followed by the formation of the AlLi (β) phase until a concentration just below 50 at.%. This phase is then in equilibrium with the Al$_2$Li$_3$ (τ) phase until 60 at.%. Further addition of Li results in a liquid phase (at 150 °C). Note that the potential for the formation of the AlLi intermetallic phase varies linearly with temperature (Huggins 1999b; Wen et al. 1979). Additional intermetallic phases also form at ambient temperature, which further complicates the phase behavior when Li is cycled with an Al electrode. Therefore, Al is not usually used as the current collector for Li deposition.

3.5.1.4 Li–Ni

The phase diagram for the Li–Ni system indicates that the solubility of Li in Ni is extremely small and no intermetallic compounds form (Predel 1997d). Ni electrodes are therefore frequently used for Li plating/stripping studies, although the high cost of Ni tends to preclude its use as a current collector in commercial batteries.

3.5.1.5 Li–Ti

The phase diagram for the Li–Ti system indicates that the solubility of Li in Ti is extremely small and no intermetallic compounds form (Bale 1989d; Predel 1997f). The two metals are almost completely immiscible, even at high temperature with liquid Li.

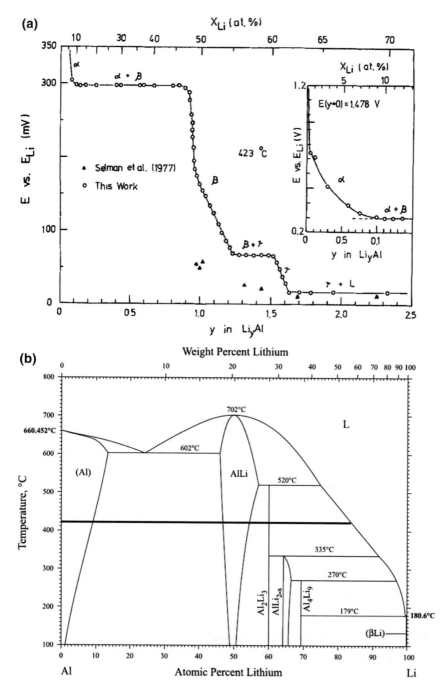

Fig. 3.31 a Coulometric titration curve for the Li–Al system at 423 °C and **b** phase diagram of the Li–Al system (the horizontal line corresponds to 423 °C. Reproduced with permission— Copyright 1970, The Electrochemical Society and 2012, Springer (Huggins 1999b; Wen et al. 1979; Okamoto 2012b)

3.5.1.6 Li–W

A phase diagram has not been reported for the Li–W system, but the solubility of Li in W is extremely small and no intermetallic compounds have been identified (Sangster and Pelton 1991a).

3.5.1.7 Li–Pt

Li reacts with Pt to form multiple intermetallic phases (Wibowo et al. 2009; Park et al. 2002; Sangster and Pelton 1991d; Lane et al. 2010b; Bronger et al. 1975, 1985). For an IL-based electrolyte containing $LiAsF_6$, after scanning to negative potential where Li–Pt alloy formation reactions occurred, four different oxidation peaks were noted when the potential was reversed and scanned positively (Wibowo et al. 2009). These peaks varied in magnitude for different temperatures (25–45 °C), indicating that the reactions progressed to a greater extent at elevated temperature and that the relative amounts of the different Li–Pt phases formed were also temperature dependent (Wibowo et al. 2009).

3.5.1.8 Li–Other Metals

No information is available regarding the phase behavior or corrosion characteristics of SS at low potentials with relevant aprotic Li^+-based electrolytes, but SS is a common cell component for Li-ion batteries and SS electrodes have been used for Li plating/stripping studies with no indications of reactions occurring between the SS and Li (Kim et al. 2013d). Other metals (such as Al and Pt) that have a complicated phase behavior with Li—with numerous intermetallic phases formed between Li and each metal—include Au (Pelton 1986a; Zeng et al. 2014), Ag (Pelton 1986b), Sn (Sangster and Bale 1998; Bailey et al. 1979), Pd (Predel 1997b; Sangster and Pelton 1992), In (Sangster and Pelton 1991c), Ga (Saint et al. 2005; Yuan et al. 2003; Sangster and Pelton 1991b), Pb (Okamoto 1993; Predel 1997e; Wang et al. 1986), Cd (van der Marel et al. 1982; Pelton 1988, 1991; Wang et al. 1986), and Zn (Okamoto 2012a; Wang et al. 1986) whereas Li has a negligible solubility in Mn (Predel 1997g), Mo (Predel 1997c, 1997h), and Rb (Bale 1989a; Predel 1997a).

3.5.2 Surface Layers and Underpotential Deposition/Stripping

Many studies of Li plating/stripping, especially those involving CV measurements, have identified one or more significant redox events prior to 0 V (vs. Li/Li$^+$) during cathodic scans which are often attributed to the underpotential depostion/stripping (UPD/UPS) of Li monolayers on Al (Li et al. 1999), Au (Aurbach 1989a; Aurbach et al. 1991b; Chang et al. 2001; Gasparotto et al. 2009; Gofer et al. 1995; Mo et al. 1996; Saito and Uosaki 2003; Wagner and Gerischer 1989; Zhuang et al. 1995; Li et al. 1986), Ag (Aurbach 1989a; Aurbach et al. 1991b), Pt (Aurbach 1989a; Aurbach et al. 1991b; Chang et al. 2001; Paddon and Compton 2007; Lee et al. 2005; Wibowo et al. 2009), Cu (Chang et al. 2001), and Ni (Li et al. 1998a; b; Wibowo et al. 2009, 2010). This phenomenon refers to the reduction of a metal cation on a solid metal surface at a potential more positive than the Nernst potential—the potential at which the metal cation will reversibly deposit on the same metal (i.e., for Li, 0 V vs. Li/Li$^+$) (Sudha and Sangaranarayanan 2002). Note that the assignment of the redox peaks to Li UPD is speculative in many of these publications with limited or no validation provided. Some of these reports mention the possibility of Li alloy formation when using Al, Au, Ag, and Pt working electrodes, and Fig. 3.29a indicates that the equilibrium alloying potentials for the different Li–Al alloys are more positive than 0 V versus Li/Li$^+$. Coulometric titration curves for Li binary systems with Zn, Cd, Pb, Sn, Sb, Bi, In, Ga, and Si reach a similar conclusion (Wang et al. 1986; Huggins 1988, 1999b; Wen and Huggins 1980, 1981), and the alloying potentials are a function of temperature (Huggins 1989b, 1999a). The diffusion of Li within these phases will be a limiting factor for determining which phases form and to what extent. Comparable titration curves are not available for Li binary systems with Al, Au, Ag, and Pt, but it is reasonable to assume that these metals also form their multiple alloy phases at potentials above 0 V versus Li/Li$^+$. Thus, this suggests that the redox peaks often attributed to UPD/UPS reactions are instead due to alloying/dealloying reactions (Tavassol et al. 2013).

Why then would redox behavior attributed to UPD be evident on Ni electrodes as well, since Ni does not alloy with Li? Fujieda et al. attributed this to the reactions of the Ni with trace water—which is reduced near 1 V (vs. Li/Li$^+$) (Fig. 3.32.) (Aurbach 1989a; Aurbach et al. 1991b)—to form an electroactive nickel hydroxide electrode surface layer (Fujieda et al. 1998). Publications by Kim et al. may also be relevant to the redox reactions noted for Cu and Ni electrodes (Kim et al. 2013d, 2011). These authors heated Cu and Ni electrodes in air. At 200 °C and above, the Cu surface formed CuO and Cu$_3$O$_4$, whereas at 300 °C and above for Ni, a NiO surface layer formed. These (nonnative) layers significantly increased the cathodic reactions occurring prior to 0 V vs. Li/Li$^+$, which is unsurprising since the oxides NiO (Wang et al. 2010; Needham et al. 2006; Liu et al. 2011a; Li et al. 2010; Kim et al. 2011; Huang et al. 2006, 2009) and Cu$_2$O/CuO (Zhang et al. 2004a; Xu et al. 2015; Wang et al. 2015b; Shu et al. 2011; Grugeon et al. 2001; Gao et al. 2004; Débart et al. 2001) have been used as anodes that undergo conversion reactions as follows:

$$Cu_2O + 2Li^+ + 2e^- \leftrightarrow 2Cu^0 + Li_2O + 2e^-$$
$$NiO + 2Li^+ + 2e^- \leftrightarrow Ni^0 + Li_2O + 2e^-$$

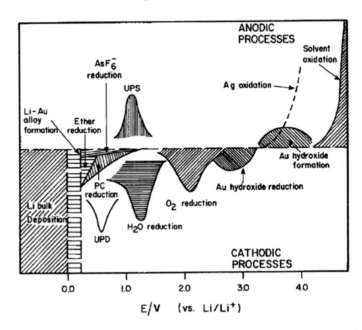

Fig. 3.32 Schematic view of typical electrochemical processes occurring in Li salt electrolytes with PC, THF, and DME on Au and Ag electrodes. Reproduced with permission—Copyright 1991, Elsevier (Aurbach et al. 1991b)

Given this, it is interesting to note that Aurbach et al. found that the reduction of dissolved O_2 occurs near 2 V (vs. Li/Li$^+$) in aprotic electrolytes (Fig. 3.32) (Aurbach 1989a; Aurbach et al. 1991b). In addition to reacting with Li$^+$ at the electrode surface to form Li_2O, this reduction of O_2 may also form metal oxide surface layers with the respective electrode metals, which then undergo subsequent conversion reactions (and the metals presumably also react with Li_2O at the metal-SEI interface). For Al and noble metal electrodes, other oxides such as Al_2O_3, PtO_2, etc., may perhaps also complicate the surface redox reactions. In addition to such considerations, aprotic solvents tend to degrade to form surface layers prior to 0 V vs. Li/Li$^+$ which further obfuscates the redox reactions occurring on varying electrodes prior to Li deposition. This confluence of reactions may explain, in part, the "incubation or initiation period" (which tends to increase with decreasing current density) that is sometimes noted to occur prior to Li plating during galvanostatic charging on metal electrodes, as well as the often observed low CE for the first cycle of Li plating/stripping (Fig. 3.33) (Brissot et al. 1999c; Ota et al. 2003, 2004a; Nishikawa et al. 2007, 2010, 2011; Nishida et al. 2013; Sano et al. 2014).

Fig. 3.33 Li plating on Ni using a 1 M IL-LiTFSI electrolyte after the start of electrolysis depending on **a** time and **b** coulomb quantity (*open circles* indicate when precipitates were first visible in the optical videos). Reproduced with permission —Copyright 2013, Elsevier (Nishida et al. 2013)

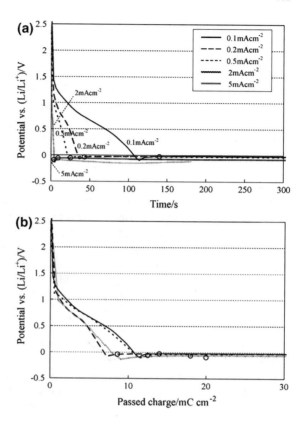

3.5.3 Surface Roughness

Dampier and Brummer noted that, when plating Li on a Ni substrate, dendritic Li accumulated as isolated material at the sharp, outer edges of the substrate (Dampier and Brummer 1977). Only a few nodules of Li were evident inside the outer border. When a second cell was constructed with the sharp edges rounded down, however, the dendrites were almost completely eliminated at the edges. Nishikawa et al. also noted that needles tended to form at the edges of a Ni electrode, whereas smaller faceted deposits instead formed in the center of the electrode, indicating that nucleation and needle growth were more rapid at the edges (Nishikawa et al. 2011). This implies that surface roughness is a significant factor for the Li deposition morphology and cycling CE.

Morigaki and Ohta (1998) noted that the Li metal surface may contain many grain boundaries, ridge lines, and flat areas. When the Li was stored in dry air, the ridge lines and grain boundaries were covered with Li_2CO_3 and Li_2O. When the Li was then stored in a 1 M $LiClO_4$-PC electrolyte for 24 h, Cl was found at the raised features, indicating that LiCl formed, whereas the solvent (PC) was degraded (perhaps forming $ROCO_2Li$) on both the raised and flat portions of the surface. After depositing Li (2 mA cm^{-2} and 0.36 C cm^{-2}) on this Li surface, particles of Li grew preferentially on the raised features and it appeared that the growth of the particles was from the base of the Li. The authors attributed this to enhanced diffusion at the raised features due to the different SEI compositions.

The mechanical modification of a Li surface using a micro-needle technique created small indentations in the surface. This increased the surface area of the Li and improved the capacity retention of batteries with these electrodes (Ryou et al. 2015). The Li was found to preferentially deposit in the striated walls of the indentations. The authors attributed this to the removal of the native surface layer on the Li substrate and to reduced current density due to the greater surface area. The results noted above from Morigaki and Ohta, however, suggest that the striated edges may be even more reactive to electrolyte impurities than the flat Li substrate surfaces. Relevant to this work is a study by Gireaud et al. that examined Li deposition on Li substrates that were pitted, cracked, or smooth (Gireaud et al. 2006). When Li was deposited on a pitted (but otherwise smooth) Li surface, dendritic Li deposits formed in the pit-like holes (comparable results were obtained for current densities of 1 and 50 mA cm^{-2}) with no Li deposited on the smooth surfaces (Figs. 3.34a) (Gireaud et al. 2006). Yoshimatsu et al. (1988) noted similar deposition behavior in pits on a Li substrate. When Li was deposited instead on the cracked Li substrate, the Li initially deposited on the crack ridge lines and these deposits then aggregated together to coat portions of the surface (Figs. 3.34b) (Gireaud et al. 2006). The separator constrained the deposits to the interiors of the pit holes. In contrast, when Li was instead plated on polished Li substrates under the same conditions, different deposits formed that had either a small grain (1 mA cm^{-2}) or free-dendritic mossy (50 mA cm^{-2}) (Fig. 3.34c) morphology all over the surface.

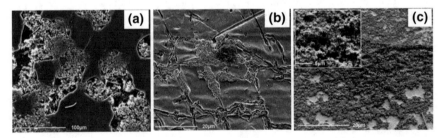

Fig. 3.34 Surface morphology after Li plating on different Li surfaces in 1 M LiPF$_6$-EC/DMC: **a** pitted Li surface; **b** cracked Li surface (10 s), **c** polished surface. *Inset* micrograph magnification showing the free-dendritic mossy deposit spread all over the surface. Reproduced with permission —Copyright 2006, Elsevier (Gireaud et al. 2006)

These factors are well reflected in the cycling characteristics of compressed Li powder anodes (Kim et al. 1988, 2006, 2007b, 2010b; Kwon et al. 2001; Park and Yoon 2003; Hong et al. 2004; Kim and Yoon 2004a, b; Chung et al. 2006; Seong et al. 2008; Kong et al. 2012; Lee et al. 2013). The rougher surface results in a higher surface area (as compared to Li foil), which reduces the effective current density on the anode, which in turn results in improved cycling characteristics (see below) (Park and Yoon 2003). Despite this, the growth in the interfacial impedance upon storage (aging—see below) was lower for the compressed Li powder than for the Li foil (Park and Yoon 2003; Kim and Yoon 2004a; Kong et al. 2012). This is likely due to the presence of a thicker native SEI film on the Li powder anode (than for the Li foil anode)—as reflected by an initially greater impedance for the powder electrode—which may suppress further reactions with the electrolyte to a greater extent (Hong et al. 2004; Kim and Yoon 2004a), as well as differences in reactivity with the electrolyte components due to the surface roughness. Rather than isolated pitting on the flat Li surface during stripping and subsequent deposition inside the pits as observed for the Li foil, the stripping of Li from the powder occurred relatively uniformly across the entire electrode, forming pits within the powder spheres. Deposition then occurred within the pits to reconstruct the compressed spheres (Kim et al. 2006, 2007b; Seong et al. 2008; Kong et al. 2012). During cycling, this resulted in lower electrolyte consumption, less dendritic growth, and reduced interphasial layer formation (Kim and Yoon 2004b; Chung et al. 2006; Kim et al. 2010b).

3.6   Influence of Charge/Discharge Profiles

3.6.1   Influence of Pulsed Plating

A coarse grain simulation model for the plating of Li metal—which accounts for the heterogeneous and nonequilibrium nature of the plating dynamics, as well as the long time- and length-scales for dendrite growth—found that dendrite formation increased with an increase in the applied overpotential, and the application of pulsed charging significantly suppressed dendrite formation at high overpotentials (Mayers et al. 2012). According to this model, dendrite growth reflects a competition between the time scales for Li^+ cation diffusion/reduction to/at the anode-SEI interface. The use of low overpotentials or short charging pulses favors cation diffusion, thus decreasing the proclivity for dendrite formation. This has been experimentally validated using Li/Li cells with a 1 M $LiPF_6$-EC/DMC (1/1 v/v) electrolyte by comparing the morphology of Li deposited by direct current (DC) and either pulse plating (PP) or reverse pulsed plating (RPP) (Fig. 3.35) (Yang et al. 2014). The PP method with short, widely spaced current pulses improved both the Li morphology (larger Li particles/nodules) and the cycling CE.

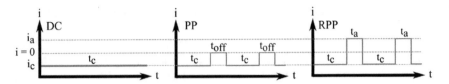

Fig. 3.35 Waveforms for DC, PP, and RPP charging, where i_a, i_c, t_a, t_c, and t_{off} are the anodic current density (mA cm^{-2}), cathodic current density (mA cm^{-2}), anodic-on time (ms), cathodic-on time (ms) and current-off time (ms), respectively. Reproduced with permission—Copyright 2014, Elsevier (Yang et al. 2014)

In contrast, the RPP method with high current density charge pulses improved the CE, but no improvement in the plating morphology was evident under the test conditions.

3.6.2   Influence of Plated Charge

Sazhin et al. examined the dependence of the CE on the plated charge amount using Li state diagrams (Fig. 3.36) (Sazhin et al. 1997). Different behavior was noted for different electrolytes. For the 1 M LiPF$_6$-EC/DEC electrolyte, the CE was relatively independent of the amount of plated charge up to a value of 0.23 C cm^{-2}, but then the CE deceased with increasing plated charge. For the 1 M LiClO$_4$-EC electrolyte, however, the CE values were much lower and declined with decreasing plating charge. This suggests that the LiPF$_6$ electrolyte passivates the Li much more effectively, but as more Li is deposited, more "dead" Li limits what may be recovered (perhaps due to longer needle-like deposits). In contrast, the LiClO$_4$ electrolyte does not effectively passivate the Li, but the buildup of the degradation products for higher plated charge may shield the deposited Li to some extent from the electrolyte, thus reducing the overall reactivity.

Ota et al. plated/stripped Li on a Ni electrode in a 1 M LiBETI-EC/THP (1/1) electrolyte using different amounts of plated charge (Ota et al. 2004a, c). The initial deposits had a fibrous shape, but with increasing applied charge, these grew into larger and larger particle/nodule-like structures (Fig. 3.37). After 50 cycles, the residual interphasial layer thicknesses (after stripping) for the plating/stripping with 0.5, 1.0, and 2.0 C cm^{-2} plated charge were 11, 26, and 49 μm, respectively (Fig. 3.37), indicating that the thickness of this layer roughly scales with the amount of plated charge. The cycling CE in the initial cycles was highest for the largest plated charge, but comparable CE values were obtained irrespective of the amount of plated charge after 10 or so cycles, and these values slowly decreased upon continuous cycling (Fig. 3.38). The larger particles may be the reason for the higher CE, but it is unclear whether the same particle deposition characteristics are retained after repeated cycling with the buildup of the interphasial layer on the electrode surface.

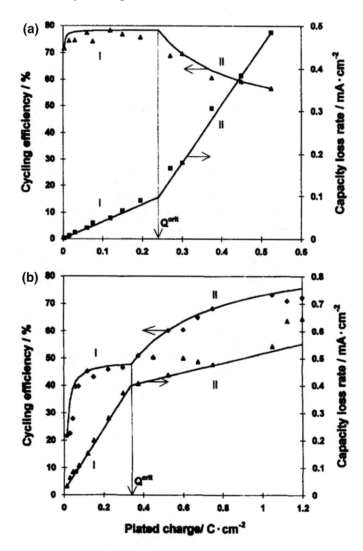

Fig. 3.36 Dependence of CE and capacity loss rate on amount of plated Li for Li plating/stripping on a SS substrate in **a** 1 M LiPF$_6$-EC/DEC and **b** 1 M LiClO$_4$-EC electrolytes. Reproduced with permission—Copyright 1997, Elsevier (Sazhin et al. 1997)

3.6.3   Influence of Plating (Charge) Current Density

The plating current density strongly affects the morphology of the deposited Li, as well as its reactivity. The use of different electrolyte compositions (i.e., different salts/solvents/additives, ILs, polymer electrolytes) often results in substantial differences in these Li plating characteristics. The wide variability in experimental

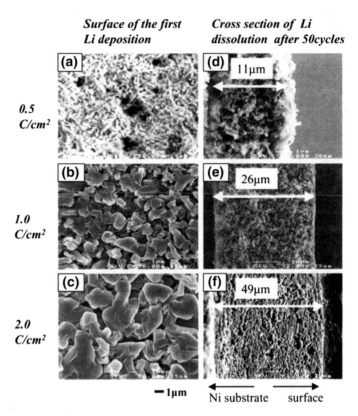

Fig. 3.37 SEM images of Li deposition for variable plated charge for Li plating on a Ni substrate in a 1 M LiBETI-EC/THP (1:1) electrolyte: **a–c** surface morphology after the initial deposition at 0.2 mA cm^{-2} and **d–f** cross-sectional morphology after 50 cycles at plating/stripping current densities of 0.2/0.6 mA cm^{-2}, respectively. Reproduced with permission—Copyright 2004, The Electrochemical Society (Ota et al. 2004a)

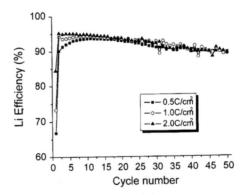

Fig. 3.38 Li cycling CE as a function of plated charge for Li plating on a Ni substrate in a 1 M LiBETI-EC/THP (1/1) electrolyte. The plating/stripping current densities were 0.2/0.6 mA cm^{-2}, respectively. Reproduced with permission—Copyright 2004, The Electrochemical Society (Ota et al. 2004a)

materials/procedures used in reported studies, however, makes it a challenge to arrive at definitive conclusions about the impact of the plating current density.

Sano et al. noted that a higher plating current density results in a higher overpotential and a corresponding larger number of nucleation sites for Li deposition (Fig. 3.39) (Nishikawa et al. 2011). Thus, they observed that the few nuclei present at low plating current density tended to grow into larger particles (for the same total charge passed), while at high current density, needle-like morphologies predominated. Ota et al. plated/stripped Li on a Ni electrode in a 1 M LiBETI-EC/THP (1/1) electrolyte at different plating current densities (with a fixed total charge for each of 0.5 C cm^{-2}) (Ota et al. 2004a, c). The cycling CE increased substantially with decreasing plating current density (Fig. 3.40). SEM images (Fig. 3.41) show that the Li had a particle-like morphology for all of the current densities, but the particles became finer with increasing rate. After 50 cycles, the residual interphasial layer thicknesses (after stripping) for the deposition at 0.2, 0.6, and 1.0 mA cm^{-2} were 11, 14, and 38 μm, respectively (Figs. 3.37 and 3.41). Thus, despite the same amount of charge for the deposition, the quantity of dead Li increased sizably with increasing plating current density—in agreement with the differing cycling CE values. This is also consistent with the observation reported by Lv et al. (2015).

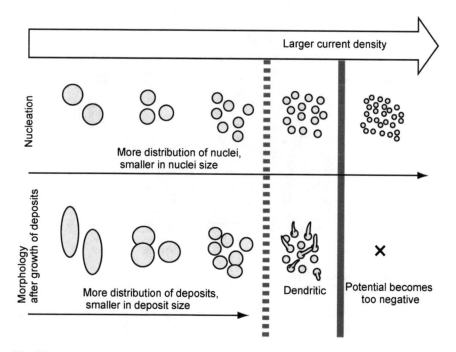

Fig. 3.39 Schematic of the effect of current density on the distribution of Li nuclei and the morphology of Li deposits. Reproduced with permission—Copyright 2014, The Electrochemical Society (Sano et al. 2014)

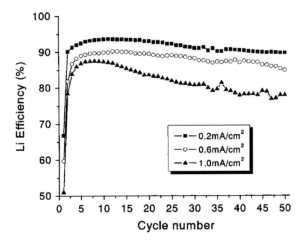

Fig. 3.40 Li cycling CE as a function of Li plating current density (0.2, 0.6 and 1.0 mA cm^{-2}) for Li plating on a Ni substrate in a 1 M LiBETI-EC/THP (1/1) electrolyte at 0.5 C cm^{-2}. The stripping current was 0.6 mA cm^{-2}. Reproduced with permission—Copyright 2004, The Electrochemical Society (Ota et al. 2004a)

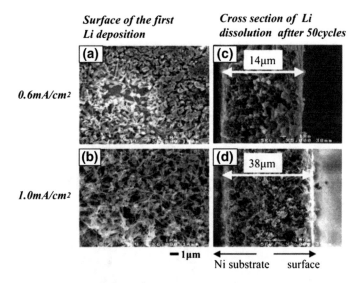

Fig. 3.41 SEM images as a function of current density for Li plating on a Ni substrate in a 1 M LiBETI-EC/THP (1/1) electrolyte at 0.5 C cm^{-2}: **a, b** surface morphology after the initial plating at 0.6 and 1.0 mA cm^{-2} and **c, d** cross-sectional morphology after 50 cycles at 0.6 and 1.0 mA cm^{-2} for plating and 0.5 mA cm^{-2} for stripping, respectively. Reproduced with permission—Copyright 2004, The Electrochemical Society (Ota et al. 2004a)

Arakawa et al. (1993) found for Li||amorphous-V_2O_5 coin cells with a LiAsF$_6$-EC/2MeTHF electrolyte that the cycle life increased with a decrease in charge current density. The cells were discharged at 3.0 mA cm^{-2} and then charged at either 0.5 or 1.5 mA cm^{-2} resulting in particle-like or needle-like deposits, respectively. In agreement with this, Aurbach et al. (2000) noted that Li||Li$_{0.3}$MnO$_2$ AA batteries with a 1 M LiAsF$_6$-DOL (stabilized with Bu$_3$N) electrolyte could be discharged at high rates for more than 300 full depth-of-discharge (DOD) charge–discharge cycles, but this required slow charging rates (<C/9). When faster charging rates were used, the cycle life decreased to less than 100 cycles. The plated Li nodular/particle-like morphology was found to decrease in size as the plating current density was increased from 0.3 to 0.75 to 1.75 mA cm^{-2}, but remained constant thereafter (i.e., for a 2.25 mA cm^{-2} rate). The higher surface area of the smaller particles led to increased reactivity with the electrolyte and thus faster electrolyte depletion (the main cause for the cell performance degeneration) (Aurbach et al. 2000; Zinigrad et al. 2004). At high rates, another failure mechanism occurred. The loss of electrolyte led to poor electrochemical contact for the cell components in the outer portion (i.e., they were less pressurized) of the AA jelly-roll batteries. Thus, the applied current created a much higher current density for the portions of the battery still in operation, which led to dendrite formation and short-circuiting of the electrodes (despite the very favorable Li plating morphology found when plated at low current density). Lv et al. (2015) also noted that the capacity of Li||nickel-cobalt-aluminum oxide (NCA) cells degraded more rapidly when charged at higher current densities (Fig. 3.42a)—in a very similar manner to that noted by Aurbach et al. (2000). This was attributed, however, to the more rapid accumulation with increasing plating current density of a thick, highly resistive interphasial layer (rather than principally to electrolyte loss). Interestingly, much of the (cathode) capacity loss was recoverable if a slow charge procedure was used after fast-charging cycles (Fig. 3.42b), suggesting that either the slower Li deposition electrically reconnected much of the dead Li, or alternatively that the Li from the underlying Li substrate became accessible through the resistive interphasial layer (Lv et al. 2015).

Dampier and Brummer (1977) found that the CE for Li cycled on a Ni electrode with a 1 M LiAsF$_6$-MA electrolyte was a strong function of the current density, with the CE decreasing rapidly below about 2 mA cm^{-2} (Fig. 3.43). Sazhin et al. reported a similar decrease in CE with decreasing current density for Li deposition on SS with a 1 M LiAsF$_6$-PC/DME (6/4) electrolyte (Dampier and Brummer 1977; Sazhin et al. 1993). In contrast, Wang et al. (1999) and Xianming et al. (2001) reported that decreasing the plating current density with LiBETI-EC/THP (1/1) and LiBETI-EC/PC/THP (3/3/4) electrolytes increased the cycling CE. These conflicting reports appear to be attributable to the differences in the reactivity of the Li with the electrolyte. As noted in the above studies, an examination of the Li morphology resulting from plating at different current densities often finds that the Li deposits are larger when grown at lower current densities with the same amount of total charge (Yang et al. 2014; Qian et al. 2015b; Arakawa et al. 1993). This seems contradictory to the observation of greater reactivity for plating at a lower current

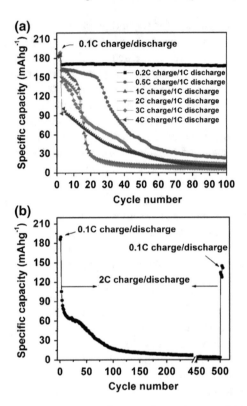

Fig. 3.42 Electrochemical evaluation of Li‖NCA cells: **a** cycling stability of cells charged at various C rates and discharged at 1C, with two formation cycles initially performed at 0.1 C for both charge and discharge and **b** cycling stability of the cell initially cycled at 2 C for 500 cycles and then switched to 0.1 C. Reproduced with permission—Copyright 2014, Wiley-VCH (Lv et al. 2015)

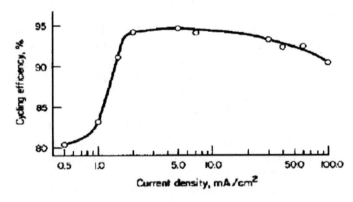

Fig. 3.43 Li cycling CE as a function of current density on a Ni substrate for a 1 M LiAsF$_6$-MA electrolyte. $Q = 0.83$ C cm^{-2} and $i_{plate} = i_{strip}$ in all cases (data for second cycle of fresh cells), where i_{plate} and i_{strip} are the Li plating and striping current density, respectively. Reproduced with permission—Copyright 1977, Elsevier (Dampier and Brummer 1977)

density that is sometimes noted, since larger Li particles will have less surface area exposed to the electrolyte. But a second important factor to consider is the time period required to achieve a specified total plated charge, which varies directly with the current density, with much longer periods required for lower current densities, thus enabling more time for side reactions to occur (Dampier and Brummer 1977).

Nishida et al. (2013) found rather different plating features when studying the growth for an unconstrained electrode (i.e., no separator and associated stack pressure). These authors monitored the Li deposition morphology on a Ni substrate with an IL electrolyte using in situ optical microscopy (Fig. 3.44). For a low current density of 0.1 mA cm^{-2}, long needle-like growth was evident. Increasing the current density to 0.2 mA cm^{-2} reduced the number of these and their extended growth did not occur until more charge had been applied. Above 0.5 mA cm^{-2}, the needle-like growth transformed to a complex structure similar to seaweed with relatively smooth deposits retained until 300 mC cm^{-2} was applied. Thus, dendrites were most prominent for both low and high plating current densities. It is unclear why these results differ from those in past studies.

The Li deposition morphology dependence on plating current density also appears to differ for studies with polymer electrolytes. Brissot et al. found that, on a Li substrate and a LiTFSI-PEO polymer electrolyte, at low current density (0.2 mA cm^{-2}), needle- and small particle-like deposits formed, whereas for higher current densities (≥ 0.7 mA cm^{-2}), the deposits were initially compact layers for

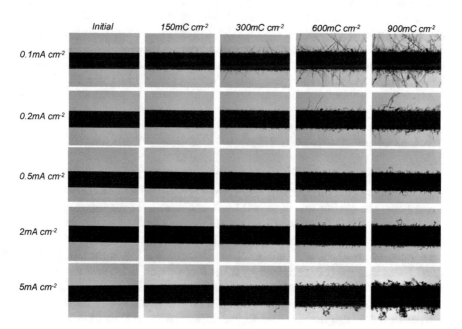

Fig. 3.44 Time transients of the morphology of plated Li on a Ni substrate in a 1 M LiTFSI-IL electrolyte as a function of current density and total current. Reproduced with permission—Copyright 2013, Elsevier (Nishida et al. 2013)

the first few cycles, but then formed tree- or bush-like deposits upon continued cycling (Brissot et al. 1998). In situ optical studies of cells with a LiTFSI-P(EO)$_n$ (n = 15–40) SPE electrolyte (cycled at 80 °C under constant-current conditions) found that at low plating current density (0.2 mA cm^{-2}), needle-like and particle-like dendrites formed. At higher plating current densities, dendrites did not form during the first polarization step (Brissot et al. 1998, 1999c). Instead, a compact Li layer formed. After several polarizations, however, dendrites did form, but now with a tree-like or bush-like morphology. New dendrites were never observed to grow on top of the dendrites produced during previous polarization. Instead they grew directly on the electrode until the entire electrode was eventually covered with them (Brissot et al. 1998). This may be due to the passivation of the dendritic Li metal surface by reactions with the electrolyte. The needle-like dendrites tended to grow either straight across the electrolyte (thus forming a short circuit with the positive electrode) or instead adopt a tortuous shape that may never induce a short circuit (Brissot et al. 1999c). Interestingly, the decrease in salt concentration near the negative electrode during cycling resulted in a compaction of the SPE, while the SPE "inflated" near the positive counter electrode (Brissot et al. 1998). This electrolyte motion is destructive to the dendrites and can lead to disconnections of dendrites that contact the positive electrode, thus resulting in variable short-circuiting behavior from cycle to cycle. The experimental conditions for the in situ measurements, however, differ significantly from a full battery and thus these observations may not be directly transferable. A very similar Li dendrite morphology versus current density trend was noted for a liquid LiTFSI-P(EO)$_n$ electrolyte (with low molecular weight PEO) (Zhang et al. 2004b). A higher plating current density, however, may result in more branching at defect sites (kinks) on the needles; thus, the tree- or bush-like Li deposits may well be generated from rapidly branching needles rather than particle-like (nodular) deposits.

3.6.4 Influence of Stripping (Discharge) Current Density

Schedlbauer et al. (2013) examined the cycling CE dependence of Li plating/stripping on a Cu electrode using 1 M LiPF$_6$- or LiDFOB-EC/DEC (3/7 wt) electrolytes. The plating was done at 0.1 mA cm^{-2}, whereas the stripping rate was varied from 0.1 to 1.0 mA cm^{-2}. Interestingly, these authors found that the CE decreased with increasing stripping current density, which contradicts what others have found, as noted below. One possible reason for this is that the stripping current density was changed (i.e., increased by 0.1 mA cm^{-2} increments) for the same cell after 10 cycles (instead of using separate cells) (Schedlbauer et al. 2013). Given that the CE is not stable for carbonate-based electrolytes and tends to decrease upon extended cycling, this may account for the disagreement. In contrast, Dampier and Brummer (1977) found that stripping the Li from a Ni electrode using a 1 M LiAsF$_6$-MA electrolyte at a low current density leads to a lower cycling CE than

Fig. 3.45 Unrecoverable
(dead) Li as a function of total
time (plating, stripping and
aging) for Li plating/stripping
on a Ni substrate with a 1 M
LiAsF$_6$-MA electrolyte.
$Q = 0.83$ C cm^{-2},
i_{plate} = plating current density
(mA cm^{-2}) and
i_{strip} = stripping current
density (mA cm^{-2}) (data for
second cycle of fresh cells).
Reproduced with permission
—Copyright 1977, Elsevier)
(Dampier and Brummer 1977)

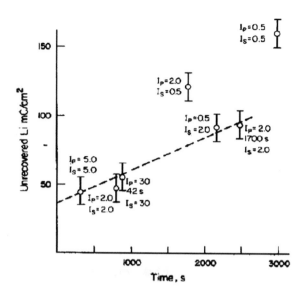

doing so at a higher current density (for Li plated at the same current density)
(Fig. 3.45). This was attributed to the exposure of newly exposed Li to the electrolyte (after disrupting the SEI as the Li was stripped) for a longer period of time
for slower discharge rates. Wang et al. (1999) noted the same trend when using a
LiBET-EC/THP (1:1 in volume ratio) electrolyte.

Laman and Brandt (1988) also found that the cycle life for Li‖MoS$_2$ batteries
with a 1 M LiAsF$_6$-EC/PC (1/1 v/v) electrolyte was strongly dependent on the
discharge rate (when a constant charge rate was used). For very high discharge
rates, the cycle life limitations were attributed to unfavorable cathode reactions. For
intermediate discharge rates, the slow buildup of high cell impedance due to
electrolyte decomposition (and the growth of the interphasial layer) eventually
limited the battery (cycle) life. For low discharge rates, however, the cycle life was
significantly shorter due to low-impedance failure (i.e., short-circuit formation by
dendrites).

Arakawa et al. (1993) indicated that comparable trends had been reported for
Li/MnO$_2$ and Li‖Li$_{0.5}$MnO$_y$ cells despite these having different cathodes, electrolytes, capacities, charge current densities, and cycling voltage ranges. These
authors extended the study of Laman and Brandt using the Li/a-V$_2$O$_5$ coin cells
with a LiAsF$_6$-EC/2MeTHF electrolyte. They also noted that the cycle life
increased with an increase in discharge current density (0.2 vs. 3.0 mA cm^{-2}).
Localized stripping occurred for the low discharge rate, while delocalized stripping
was evident for the high rate. This was attributed to variations in the resistance in
the SEI. For the low rate, the induced overpotential may not be sufficient to strip the
Li from the more resistive parts of the SEI, but the higher overpotential at higher
rates may drive stripping from a greater portion of the Li surface. When the cells

were charged at 0.4 mA cm^{-2} after discharge, needle-like deposits grew in the cell discharged at 0.2 mA cm^{-2}, while particle-like deposits formed instead in the cell discharged at 3.0 mA cm^{-2}. This difference occurred because the deposited Li was localized where the Li was previously stripped during the discharge step, in accord with what was reported above for surface roughness.

Zheng et al. (2015) have recently reported a similar improvement in the cycling performance with increasing discharge rates for Li‖NMC cells with a 1 M LiPF$_6$-EC/DMC (1/2 v/v) electrolyte using a fixed charge rate. A substantial interphasial layer formed at the lower discharge rates that was largely absent for the cells discharged at higher rates. This was attributed to a mechanism in which the higher concentration of Li$^+$ cations near the Li surface from the higher stripping rate created a transient layer, which created a flexible SEI layer composed of poly (ethylene carbonate) and other organic/inorganic Li salts.

3.7 Effect of Rest/Storage Time

It was noted above that aging the plated Li results in a decrease in the amount of Li that may be subsequently stripped (i.e., an increase in the amount of dead Li). The high chemical reactivity of Li often results in continuous reactions with the electrolyte components (sometimes referred to as a self-discharge rate, Li corrosion or Li encapsulation), even when no electrochemical reactions occur, unless a stable, compact passivation layer is formed that limits or prevents such contact—as has been demonstrated by Aurbach and Mengeritsky et al. using tailored formulations with LiAsF$_6$-DOL stabilized with Bu$_3$N (Mengeritsky et al. 1996b; Gofer et al. 1992).

Dampier and Brummer (1977) conducted an evaluation of the effect of storage/aging on the self-discharge rates of electrodeposited Li in 1 M LiAsF$_6$-MA electrolytes. The rate of self-discharge was highest just after the deposit was plated, and subsequently decreased during storage (Fig. 3.46). Similar effects were noted by Besenhard, Rauh, and Brummer using LiClO$_4$-PC and LiClO$_4$-MA electrolytes

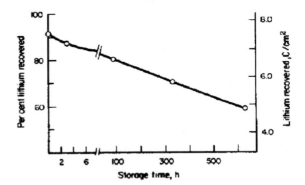

Fig. 3.46 Self-discharge rate of Li deposited on Ni substrates in a 1 M LiAsF$_6$-MA electrolyte and aged over a 26 day period. $Q = 0.83$ C cm^{-2} and $i_{plate} = 5.0$ mA cm^{-2}. Reproduced with permission —Copyright 1977, Elsevier (Bieker et al. 2015)

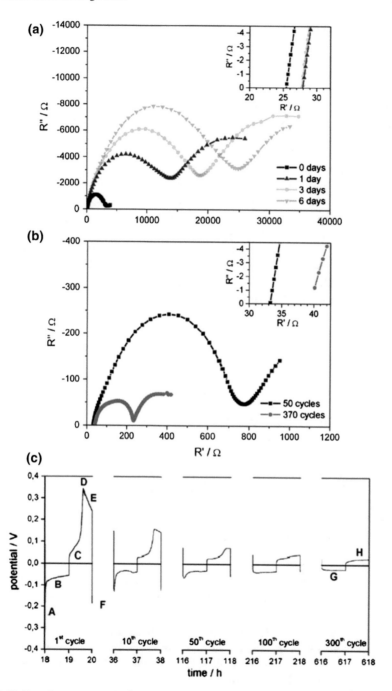

Fig. 3.47 Impedance spectra of Li‖Li cells with a 1 M LiPF$_6$-EC/DEC (3:7) electrolyte: **a** after 0, 1, 3, and 6 days at OCP and **b** after 50 and 370 plating/stripping cycles at 0.1 mA cm^{-2}. **c** Selected overpotential profiles for the cells. Reproduced with permission—Copyright 2015, The Royal Society of Chemistry (Ota et al. 2004a)

(Rauh and Brummer 1977a, b; Besenhard 1977). Notably, Dampier and Brummer determined that, for a wide range of current densities, the total cycling time was what principally determined the amount of unrecoverable Li—i.e., the same results were obtained when the cycling time was increased by the inclusion of a time delay after the plating, as was noted for an increase in the plating (or stripping) time by decreasing the current density.

Bieker et al. (2015) have also recently conducted an extensive exploration of the effect of rest/storage time on the Li electrode. Impedance spectra of Li∥Li cells with a 1 M $LiPF_6$-EC/DEC (3/7) electrolyte after 0, 1, 3, and 6 days at open circuit potential (OCP) indicate that the interphasial layer continues to grow in thickness, resulting in a substantially resistive layer (Fig. 3.47a). This creates a high overpotential when the cell is cycled (Fig. 3.47c), but this layer is disrupted upon extended cycling, leading to a lower interfacial impedance and low overpotential (Figs. 3.47b and 3.47c). Similar results have been reported by Orsini et al. (2000). The variability in the overpotential upon cycling (Fig. 2.20), which is commonly noted during continuous Li plating/stripping cycling, may therefore be due to the intermittent disruption and growth of the interphasial layer where the Li is being plated (i.e., surface reorganization) (Best et al. 2010; Bhatt et al. 2010, 2012; Basile et al. 2013). It is important to note that this electrode surface resistance is far larger than that of the electrolyte resistance, indicating that transport through the SEI is likely to be the prime determinant for the Li overpotential for the plating/stripping rather than bulk electrolyte conductivity.

3.8 Effect of Temperature

As noted above, temperature in some cases has a strong influence on the Li plating morphology. This may be due to the link between Li^0 adatom mobility on the growing Li surface, as well as the mobility of the Li^+ cations within the SEI layer (as well as the bulk electrolyte). This has not been extensively explored in the scientific literature, but results of some relevant reports follow.

Studies that examined the CE and resistance for Li plated at different temperatures found differences in the trends that were electrolyte solvent dependent. When a 1 M $LiPF_6$-PC/DMC electrolyte was used, the CE improved for lower temperature cycling, as well as when the Li was initially plated for 10 cycles at either 0 or −20 °C followed by cycling at 25 °C (relative to the Li plated solely at 25 °C) (Ishikawa et al. 1999a). The CE was also found to increase upon decreasing the temperature from 25 to 10 to 0 °C for a 1 M $LiCF_3SO_3$-EC/DMC electrolyte (Matsuda et al. 1997). In contrast, when a 1 M $LiPF_6$-PC/2MeTHF electrolyte was used, the CE decreased with decreasing temperature (Ishikawa et al. 1999a). These differences were attributed to the interphasial layer, which had a relatively low

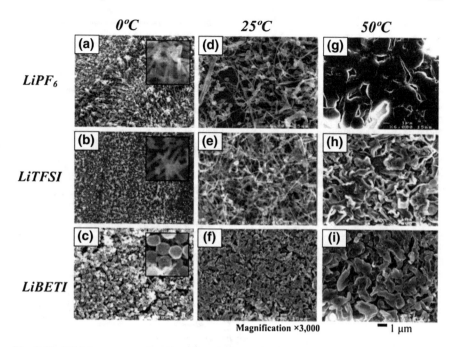

Fig. 3.48 SEM images as a function of temperature of Li deposited on a Ni substrate in 1 M LiX-EC/THF (1:1) electrolytes containing LiPF$_6$, LiTFSI and LiBETI, respectively: **a–c** 0 °C, **d–f** 25 °C and **g–i** 50 °C. The plating current was 0.6 mA cm^{-2} up to 0.5 C cm^{-2}. Reproduced with permission—Copyright 2004, The Electrochemical Society (Ota et al. 2004a)

resistance for the DMC-containing electrolyte, but a much more resistive layer for the 2MeTHF electrolyte.

To examine the effect of temperature on the plating morphology, Ota et al. (2004a) deposited Li on Ni from 1 M LiX-EC/THF (1/1) electrolytes with either LiPF$_6$, LiTFSI or LiBETI at 0, 25, or 50 °C (Fig. 3.48). Notably, at 50 °C the Li particles were larger and somewhat flat, whereas at 0 °C the Li particles were much smaller for the electrolyte with LiBETI, but were dendritic fibers for the electrolytes with LiPF$_6$ and LiTFSI. Impedance spectroscopy was used to examine the resistance of the deposited Li. The Li deposited at 50 °C in the LiPF$_6$ electrolyte had a higher resistance and lower CE than the Li deposited in the LiTFSI and LiBETI electrolytes (Fig. 3.49). This may, in part, be due to the larger particle size and thus lower surface area of the former.

On switching to 25 °C, the resistance of the Li from LiPF$_6$ remained higher, but the value decreased upon continuous cycling and the CE improved (Fig. 3.50a). When the Li was instead initially deposited and cycled at 0 °C, the resistance and

Fig. 3.49 Impedance plots after the initial Li plating as a function of electrolyte temperature in 1 M LiX-EC/THF (1/1) electrolytes containing LiPF$_6$, LiTFSI, and LiBETI, respectively: **a** 0 °C, **b** 25 °C and **c** 50 °C. The plating current was 0.6 mA cm^{-2} up to 0.5 C cm^{-2}. Reproduced with permission—Copyright 2004, The Electrochemical Society (Ota et al. 2004a)

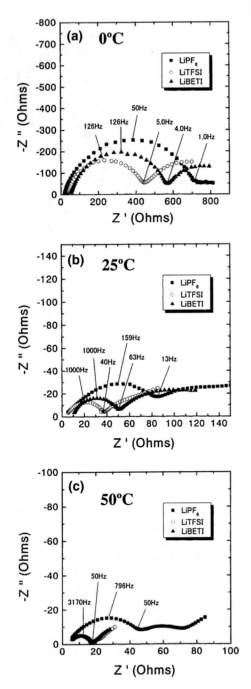

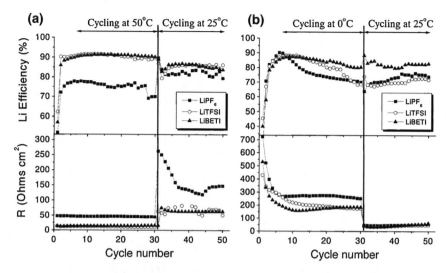

Fig. 3.50 Li cycling CE and change in resistance for Li plating/stripping on a Ni substrate in 1 M LiX-EC/THF (1:1) electrolytes containing LiPF$_6$, LiTFSI, and LiBETI, respectively, with the temperature changed after 30 cycles **a** from 50 to 25 °C or **b** from 0 to 25 °C. The current densities were 0.6/0.6 mA cm^{-2} (0.5 C cm^{-2}) for the plating/stripping, respectively. Reproduced with permission—Copyright 2004, The Electrochemical Society (Ota et al. 2004a)

CE were initially similar for the electrolytes, but the values for LiPF$_6$ became higher and lower, respectively, upon continuous cycling (Fig. 3.50b). When the temperature was switched to 25 °C, however, the resistance values became similar for all of the electrolytes and the resistance values at 25 °C for the Li plated from the LiPF$_6$ electrolyte at 0 °C were much lower than those for the Li plated from the same electrolyte at 50 °C. Note, however, that this does not affect the cycling CE substantially. These results are consistent with the improved cycling performance noted with 1 M LiBETI-PC and 1 M LiCF$_3$SO$_3$-PC electrolytes at elevated (80°C) temperature (Mogi et al. 2001, 2002a, b). In general, the Li morphology became more particle/nodule-like and larger, with a corresponding improvement in CE during cycling at higher temperature (Table 3.2).

Nishida et al. (2013) used in situ optical microscopy to examine the influence of temperature on Li plated from a 1 M LiTFSI-IL electrolyte. With increasing temperature and plated charge, the growth of long needle-like deposits became enhanced, and these thickened and aggregated together into larger clusters (Fig. 3.51). Thus, the influence of temperature is complicated by other factors (e.g., the SEI) that also strongly impact the Li morphology. If needles form, then these grow more rapidly at elevated temperature, resulting in dendritic Li. If particles/nodules form instead, then these grow larger with increasing temperature, resulting in a less fibrous structure for the deposit.

Table 3.2 Effect of electrolyte temperature on Li deposition in 1 M LiX-EC/THF (1/1) electrolytes (Ota et al. 2004a)

| Salt | Temperature | Li cycle CE (%) | | Cycle stability | Morphology | Thickness after 50 cycles (μm) |
		Avg. 1–20 cycles	Avg. 1–50 cycles			
LiPF$_6$	0	78	74	Bad	Dendritic	29
	25	80	80	Good	Particle/dendritic	24
	50	81	79	Good	Large particles	41
LiTFSI	0	81	72	Bad	Dendritic	26
	25	87	83	Bad	Dendritic	29
	50	89	89	Very good	Large particles	31
LiBETI	0	83	81	Good	Fine particles	24
	25	87	88	Very good	Particles	15
	50	89	89	Very good	Large particles	24

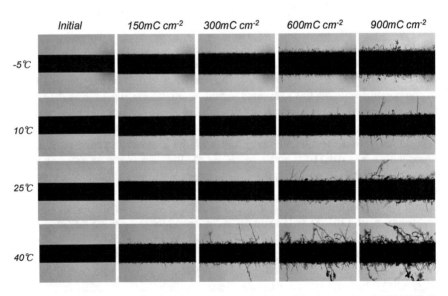

Fig. 3.51 Time transients of the morphology of plated Li on a Ni substrate in a 1 M LiTFSI-IL electrolyte as a function of temperature and total current. Reproduced with permission—Copyright 2013, Elsevier (Nishida et al. 2013)

3.9   Effect of Stack Pressure

Stack pressure has a dramatic influence on the compactness and porosity of the plated Li (Fig. 3.52) and thus on the exposure of the Li to the electrolyte. A reduction in the Li surface area results in less overall SEI formation and better electrical contact with the current collector. This, in turn, results in an improved cycling CE and possibly a reduced propensity for dendrites to contact the cathodes.

Hirai et al. noted improvements in the Li cycling CE with increasing stack pressure using a coin-type cell pressurized by a hydraulic hand press using a variety of electrolytes containing different mixtures of solvents (EC, PC, 2MeTHF) and salts (LiAsF$_6$, LiPF$_6$, LiCF$_3$SO$_3$) (Hirai et al. 1994a, b). The cells contained a cathode (70 wt% a-V$_2$O$_5$-P$_2$O$_5$, 25 wt% acetylene black, and 5 wt% polytetrafluoroethylene [PTFE]) and a Li disk anode with a Celgard 5511 separator (for PC-based electrolytes) or Celgard 2502 separator (for other electrolytes). The Li cycling CE was evaluated using an "accelerated test" in which the cycling capacity

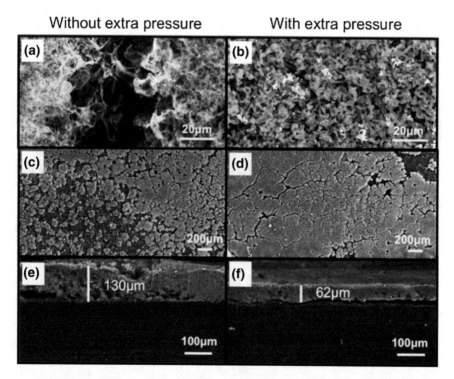

Fig. 3.52 SEM images of Li microstructures formed under different external pressures after charging at a current density of 1.1 mA cm^{-2} for 240 min for a cell with a Celgard separator and 1 M LiPF$_6$-EC/DEC: **a**, **b** surface microstructure, **c**, **d** surface image, and **e**, **f** cross-sectional image. Reproduced with permission—Copyright 2015, The American Chemical Society (Chang et al. 2015)

of the cell was limited by the Li content (Hirai et al. 1994a). Interestingly, when the electrolyte salt and salt concentration were kept constant (i.e., 1.5 M $LiAsF_6$), a substantial difference in the efficiency was found for different solvents. The cell with $LiAsF_6$-EC/2MeTHF had a large increase in efficiency with increasing pressure, as did the cell with $LiAsF_6$-2MeTHF, although to a lesser extent. A much smaller effect was seen for the cells with the $LiAsF_6$-EC/PC and $LiAsF_6$-PC electrolytes, but this may be attributable to the use of a different separator. For the cells with the same solvent mixture (EC/2MeTHF), but different (1.5 M) Li salts, a substantial improvement with increasing pressure was found for the cell with the $LiAsF_6$-based electrolyte, whereas a pressure increase in the cells with the electrolytes containing $LiPF_6$ and $LiCF_3SO_3$ was much less beneficial to the cycling CE (Hirai et al. 1994a).

Swagelok cells equipped with springs of differing stiffness have been used by Gireaud et al. (2006) to explore the influence of stack pressure on Li plating. The working electrode was an SUS 16L SS plate and the electrolyte was a 1 M $LiPF_6$-EC/DMC electrolyte. Increasing pressure resulted in more compact Li plating, which in turn resulted in a higher CE for the cycling. Similarly, Eweka et al. (1997) used a spring and screwable cell lid to apply stack pressure and found that a notable improvement in the cycling CE was evident when using a 1 M $LiAsF_6$-EC/PC electrolyte. Without the stack pressure, the initial cycling CE was below 70 %, and this rose to approximately 90 % with the applied pressure, but the effect was not persistent, as the CE began to rapidly decline after 25–35 cycles.

Wilkinson and Wainwright reported highly informative studies using a force-displacement apparatus to measure the electrode thickness as a function of pressure during cycling (Wilkinson et al. 1991; Wilkinson and Wainwright 1993). For Li||Li_xMoS_2 flat plate cells with a 1 M $LiAsF_6$-PC electrolyte, there was a strong correlation between the pressure and the Li cycling CE, with a marked improvement at high pressure (especially above a critical pressure of 60–80 psi). SEM images of the Li electrodes indicated that a very different Li plating morphology occurred below and above this critical pressure. At low pressure, a porous Li deposit formed with a variety of differently shaped crystals (needle-like and lump-shaped). At 200 psi, however, the deposit instead consisted of closed-packed columns extending from the uncycled Li substrate (Wilkinson et al. 1991). The improved cycling CE noted at higher pressure was attributed to the decreased surface area of the compact Li metal, and thus decreased rate of reaction with the electrolyte. A large change in the electrode stack thickness was noted during cycling since the volume of the plated Li metal is greater than the volume decrease of the cathode material due to delithiation (Fig. 3.53a) (Wilkinson et al. 1991). For two cells at different pressures, in addition to this cyclic variation in thickness from the Li metal plating/stripping, there is a net increase in the stack thickness for both cells, but substantially more for the cell with the lower applied pressure (Fig. 3.53b) (Wilkinson et al. 1991) due to the more rapid accumulation of a thick interphasial layer (Wilkinson and Wainwright 1993). The average net growth per cycle varied with the electrolyte used, as might be expected from differences in electrolyte component reactivity with the Li metal, with a 0.75 M $LiAsF_6$-sulfolane/tri-glyme

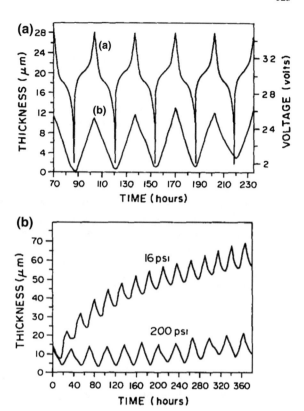

Fig. 3.53 Thickness and voltage of a Li‖Li$_x$MoO$_2$ flat plate cell containing 1 M LiAsF$_6$-PC/(85/15 v/v) EC: **a** cell cycled at 200 psi stack pressure, 0.30 mA cm^{-2}, 2.0–3.5 V, and 21 °C (initial thickness is arbitrary) and **b** cells cycled under the same conditions, but two different pressures. Reproduced with permission—Copyright 1991, Elsevier (Wilkinson et al. 1991)

(G3) (50/50 v/v) electrolyte resulting in significantly more growth than a 1 M LiAsF$_6$-PC/EC (85/15 v/v) electrolyte.

Wilkinson and Wainwright found that the Li deposits formed when cycling at low pressure (<60 psi) tended to penetrate the 25 μm thick separator after only a few cycles, resulting in a soft short (or shunt) for the charging current (Wilkinson et al. 1991). Thus, a charge imbalance occurred in which excess charging consumed some of the input current intended for recharging (Li plating). This charge imbalance was not evident for high pressure cells (100–300 psi) with the more compact Li deposit. Interestingly, if the initial cycling was done at very high pressure (e.g., 500 psi), then a shunt usually occurred on the first recharge. Once this shunt formed, no further electrode thickness growth was noted because all of the current was passed through the shunt (i.e., no Li deposition occurred). However, if the cycling was instead initially done at a lower pressure (100–200 psi) before the very high pressure was applied, no shunting occurred (Wilkinson et al. 1991). No apparent damage was evident for the microporous separator for either case, and this effect was not well understood.

Except for the case of small diameter coin cells and perhaps some prismatic or pouch cell designs, it is impractical to design cells with a fixed stack pressure, since the majority of cells are based upon constant-volume designs such as spirally wound cylindrical cells. For such cells, however, the Li layer growth during cycling will eventually build up the pressure, but the porous (dendritic) Li plating that initially occurs at lower pressure during the initial cycling may create detrimental effects for the subsequent cycling (Wilkinson et al. 1991). In addition, the pressure may vary significantly within the cell as a function of radius. Electrode stack pressures greater than 1000 psi have been found at the core of spirally wound AA size cells, which resulted in separator damage after long-term cycling (Wainwright and Shimizu 1991). This affects all of the cell components. For example, although Celgard 2500 is a very incompressible polymer membrane, a wide disparity was found in the thickness and Gurley values (Arora and Zhang 2004) (i.e., air permeability) after cell cycling (Wainwright and Shimizu 1991). The electrode pressure can be controlled to some extent by regulating the foil tensions and adjusting the fit of the electrode assembly within the cell can, and this, in turn, is a determinant for the morphology of the plated Li metal during charging (Wilkinson et al. 1991). It is also possible that pressure differentials within such batteries result in Li redistribution during cycling, which would greatly affect the cycling performance and life of the cell (Ferrese and Newman 2014). The application of force to improve the electrochemical performance has been patented (Scordilis-Kelley et al. 2015).

3.10 Summary

The Li cycling CE is strongly dependent upon the Li morphology, which is governed in large part by the SEI composition. The SEI is typically a multilayer structure on the surface of the Li, with the species abutting the Li being the most reduced, and those close to the electrolyte interface often having organic constituents. Additives/impurities are typically the most reactive electrolyte components and are thus reduced before other SEI components. Thus, even ppm amounts of species such as H_2O, O_2, and CO_2 can drastically change the SEI. Solvent reduction species are often prominent for reactive solvents (e.g., esters and carbonates), while anion reactions can be more influential when more stable solvents (e.g., ethers) are present. The electrolyte salt concentration can have a major impact on the CE for highly concentrated electrolytes, but this is strongly influenced by the solvent and Li salt employed. The replacement of aprotic solvents with ILs reduces, but does not fully prevent, formation of Li dendrites. ILs (and inorganic fillers) may also be added to SPEs to improve their Li plating/stripping attributes and reduce the tendency for dendrites to form. Block copolymer electrolytes consisting of polystyrene and poly(ethylene oxide) have been extensively evaluated in an effort to prevent dendrite growth through these rigid electrolytes. While these do hinder dendrite infiltration, they do not ultimately prevent short-circuiting of cells due to dendrites. This is also the case for solid inorganic (crystalline or glassy) separators.

The reactivity of Li with electrolyte components leading to SEI formation results in significant differences in the Li electrodeposition morphology from that of other metals. Rather than forming tree-like dendrites, Li tends to instead form either kinked needles (one-dimensional growth) and/or nodules (three-dimensional growth). Branching of the needles and/or aggregation of the needles/nodules leads to more complicated constructs that are often also referred to as dendrites. The substrate on which the Li is plated is a significant factor governing the deposition. Reactivity with electrolyte components (including impurities) leads to organic, inorganic (salt), and metal oxide surface film formation, which influences (and possibly dominates) ion transport to and reactivity with the deposition surfaces. Substrate surface roughness (pits, ridge lines, etc.) also dramatically affects where and how Li deposits. For some metal substrates, alloy formation further complicates the interpretation of the electrochemical redox reactions occurring. The reaction conditions (pulsed plating, plated charge, and plating/stripping current density), as well as aging (storage) of the deposited Li, affect not only the deposition morphology, but also the amount of Li that may be electrochemically stripped due to continuous electrochemical/chemical side reactions that result in dead (electrochemically inaccessible) Li—the extent of which is often dictated by the surface area and exposure time between the deposited Li and electrolyte. This, in turn, creates the highly resistive interphasial (mossy) Li layer that typically builds up on the Li surface after repeated cycling. This layer—rather than short-circuiting from Li dendrites—is often what limits the cell capacity and lifetime of cells using Li metal anodes. Plating at elevated temperature tends to facilitate three-dimensional growth of the Li deposits, whereas plating at subambient temperature often results instead in one-dimensional growth. Finally, the application of stack pressure generally results in compression of the Li deposit and superior plating/stripping behavior due to both lower side reaction rates (with the less porous Li) and less dead Li due to better electrical contact within the deposited Li.

References

Abraham KM (1985) Recent developments in secondary lithium battery technology. J Power Sour 14:179–191

Abraham KM, Goldman JL (1983) The use of the reactive ether, Tetrahydrofuran (THF), in rechargeable lithium cells. J Power Sour 9:239–245

Abraham KM, Goldman JL, Natwig DL (1982) Characterization of ether electrolytes for rechargeable lithium cells. J Electrochem Soc 129(11):2404–2409

Abraham KM, Foos JS, Goldman JL (1984) Long cycle life secondary lithium cells utilizing tetrahydrofuran. J Electrochem Soc 131(9):2197–2199

Abraham KM, Pasquariello DM, Martin FJ (1986) Mixed ether electrolytes for secondary lithium batteries with improved low temperature performance. J Electrochem Soc 133(4):661–666

Appetecchi GB, Passerini S (2002) Poly(ethylene oxide)-LiN(SO2CF2CF3)2 polymer electrolytes. J Electrochem Soc 149(7):A891–A897. doi:10.1149/1.1483098

Appetecchi GB, Croce F, Dautzenberq G, Mastragostino M, Ronci F, Scrosati B, Soavi F, Zanelli A, Alessandrini F, Prosini PP (1998) Composite polymer electrolytes with improved lithium metal electrode interfacial properties I. Electrochemical properties of dry PEO-LiX systems. J Electrochem Soc 145(12):4126–4132

Appetecchi GB, Croce F, Ronci F, Scrosati B, Alessandrini F, Carewska M, Prosini PP (1999) Electrochemical characterization of a composite polymer electrolyte with improved lithium metal electrode interfacial properties. Ionics 5:59–63

Appetecchi GB, Scaccia S, Passerini S (2000) Investigation on the stability of the lithium-polymer electrolyte interface. J Electrochem Soc 147(12):4448–4452

Appetecchi GB, Kim GT, Montanino M, Carewska M, Marcilla R, Mecerreyes D, De Meatza I (2010) Ternary polymer electrolytes containing pyrrolidinium-based polymeric ionic liquids for lithium batteries. J Power Sour 195(11):3668–3675. doi:10.1016/j.jpowsour.2009.11.146

Appetecchi GB, Kim GT, Montanino M, Alessandrini F, Passerini S (2011) Room temperature lithium polymer batteries based on ionic liquids. J Power Sour 196(16):6703–6709. doi:10.1016/j.jpowsour.2010.11.070

Arakawa M, Tobishima S-I, Nemoto Y, Ichimura M (1993) Lithium electrode cycleability and morphology dependence on current density. J Power Sour 43–44:27–35

Arakawa M, Tobishima S, Hirai T, Yamaki J (1999) Effect of purification of 2-Methyltetrahydrofuran/Ethylene carbonate mixed solvent electrolytes on cyclability of lithium metal anodes for rechargeable cells. J Appl Electrochem 29:1191–1196

Arie AA, Lee JK (2011) Electrochemical characteristics of lithium metal anodes with diamond like carbon film coating layer. Diam Relat Mater 20(3):403–408. doi:10.1016/j.diamond.2011.01.040

Armstrong RD, Brown OR, Ram RP, Tuck CD (1989) Lithium electrodes based upon aluminum and alloy substrates i. impedance measurements on aluminum. J Power Sour 28:259–267

Arora P, Zhang Z (2004) Battery separators. Chem Rev 104:4419–4462

Aurbach D (1989a) The electrochemical behavior of lithium salt solutions of γ-butyrolactone with noble metal electrodes. J Electrochem Soc 136(4):906–913

Aurbach D (1989b) Identification of surface films formed on lithium surfaces in γ-butyrolactone solutions. 1. uncontaminated solutions. J Electrochem Soc 136(6):1606–1610

Aurbach D (1999) The electrochemical behavior of active metal electrodes in nonaqueous solutions. Nonaqueous Electrochem (Ed Aurbach, D):289–411

Aurbach D, Chusid (Youngman) O (1993) In situ FTIR spectroelectrochemical studies of surface films formed on Li and nonactive electrodes at low potentials in Li salt solutions containing CO_2. J Electrochem Soc 140 (11):L155–L157

Aurbach D, Gofer Y (1991) The behavior of lithium electrodes in mixtures of alkyl carbonates and ethers. J Electrochem Soc 138(12):3529–3536

Aurbach D, Gottlieb H (1989) The electrochemical behavior of selected polar aprotic solvents. Electrochim Acta 34(2):141–156

Aurbach D, Granot E (1997) The study of electrolyte solutions based on solvents from the "Glyme" family (linear polyethers) for secondary Li battery systems. Electrochim Acta 42 (4):697–718

Aurbach D, Moshkovich M (1998) A study of lithium deposition-dissolution processes in a few selected electrolyte solutions by electrochemical quartz crystal microbalance. J Electrochem Soc 145(8):2629–2639

Aurbach D, Daroux ML, Faguy PW, Yeager E (1987) Identification of surface films formed on lithium in propylene carbonate solutions. J Electrochem Soc 134(7):1611–1620

Aurbach D, Daroux ML, Faguy PW, Yeager E (1988) Identification of surface films formed on lithium in dimethoxyethane and tetrahydrofuran solutions. J Electrochem Soc 135(8):1863–1871

Aurbach D, Gofer Y, Langzam J (1989) The correlation between surface chemistry, surface morphology, and cycling efficiency of lithium electrodes in a few polar aprotic systems. J Electrochem Soc 136(11):3198–3205

Aurbach D, Youngman O, Dan P (1990a) The electrochemical behavior of 1,3-Dioxolane-LiClO$_4$ solutions—II Contaminated solutions. Electrochim Acta 35(3):639–655

Aurbach D, Youngman O, Gofer Y, Meitav A (1990b) The electrochemical behavior of 1,3-Dioxolane-LiClO$_4$ solutions—I. Uncontaminated solutions. Electrochim Acta 35(3):625–638

Aurbach D, Skaletsky R, Gofer Y (1991a) The electrochemical behavior of calcium electrodes in a few organic electrolytes. J Electrochem Soc 138(12):3536–3545

Aurbach D, Daroux M, Faguy P, Yeager E (1991b) The electrochemistry of noble metal electrodes in aprotic organic solvents containing lithium salts. J Electroanal Chem 297:225–244

Aurbach D, Gofer Y, Ben-Zion M, Aped P (1992) The behaviour of lithium electrodes in propylene and ethylene carbonate: the major factors that influence Li cycling efficiency. J Electroanal Chem 339:451–471. doi:http://dx.doi.org/10.1016/0022-0728(92)80467-I

Aurbach D, Daroux M, McDougall G, Yeager EB (1993) Spectroscopic studies of lithium in an ultrahigh vacuum system. J Electroanal Chem 358:63–76. doi:http://dx.doi.org/10.1016/0022-0728(93)80431-G

Aurbach D, Ein-Ely Y, Zaban A (1994a) The surface chemistry of lithium electrodes in alkyl carbonate solutions. J Electrochem Soc 141(1):L1–L3

Aurbach D, Weissman I, Zaban A, Chusid O (1994b) Correlation between surface chemistry, morphology, cycling efficiency and interfacial properties of Li electrodes in solutions containing different Li salts. Electrochim Acta 39:51–71

Aurbach D, Zaban A, Gofer Y, Abramson O, Ben-Zion M (1995a) Studies of Li anodes in the electrolyte system 2Me-THF/THF/Me-Furan/LiAsF6. J Electrochem Soc 142(3):687–696

Aurbach D, Zaban A, Schechter A, Ein-Eli Y, Zinigrad E, Markovsky B (1995b) The study of electrolyte solutions based on ethylene and diethyl carbonates for rechargeable Li Batteries. I. Li metal anodes. J Electrochem Soc 142(9):2873–2882

Aurbach D, Zaban A, Gofer Y, Ely YE, Weissman I, Chusid O, Abramson O (1995c) Recent studies of the lithium-liquid electrolyte interface. Electrochemical, morphological and spectral studies of a few important systems. J Power Sour 54:76–84

Aurbach D, Markovsky B, Shechter A, Ein-Eli Y (1996) A comparative study of synthetic graphite and Li electrodes in electrolyte solutions based on ethylene carbonate-dimethyl carbonate mixtures. J Electrochem Soc 143(12):3809–3820

Aurbach D, Zaban A, Ein-Eli Y, Weissman I, Chusid O, Markovsky B, Levi M, Levi E, Schechter A, Granot E (1997) Recent studies on the correlation between surface chemistry, morphology, three-dimensional structures and performance of Li and Li-C intercalation anodes in several important electrolyte systems. J Power Sour 68:91–98

Aurbach D, Zinigrad E, Teller H, Dan P (2000) Factors which limit the cycle life of rechargeable lithium (metal) batteries. J Electrochem Soc 147:1274–1279

Aurbach D, Zinigrad E, Cohen Y, Teller H (2002a) A short review of failure mechanisms of lithium metal and lithiated graphite anodes in liquid electrolyte solutions. Solid State Ionics 148:405–416

Aurbach D, Zinigrad E, Teller H, Cohen Y, Salitra G, Yamin H, Dan P, Elster E (2002b) Attempts to improve the behavior of Li electrodes in rechargeable lithium batteries. J Electrochem Soc 149 (10):A1267–A1277. doi:10.1149/1.1502684

Baek S-W, Lee J-M, Kim TY, Song M-S, Park Y (2014) Garnet related lithium ion conductor processed by spark plasma sintering for all solid state batteries. J Power Sour 249:197–206. doi:10.1016/j.jpowsour.2013.10.089

Bailey DM, Skelton WH, Smith JF (1979) Lithium-Tin phase relationship between Li_7Sn_2 and LiSn. J Less-Common Metal 64:233–240

Balducci A, Jeong SS, Kim GT, Passerini S, Winter M, Schmuck M, Appetecchi GB, Marcilla R, Mecerreyes D, Barsukov V, Khomenko V, Cantero I, De Meatza I, Holzapfel M, Tran N (2011) Development of safe, green and high performance ionic liquids-based batteries (ILLIBATT Project). J Power Sour 196 (22):9719–9730. doi:10.1016/j.jpowsour.2011.07.058

Bale CW (1989a) The Li–Rb (Lithium–Rubidium) system. Bull Alloy Phase Diag 10(3):268–269

Bale CW (1989b) The Cs–Li (Cesium–Lithium) system. Bull Alloy Phase Diag 10(3):232–233

Bale CW (1989c) The Li–Na (Lithium–Sodium) system. Bull Alloy Phase Diag 10(3):265–268

Bale CW (1989d) The Li–Ti (Lithium–Titanium) system. Bull Alloy Phase Diag 10(2):135–138

Baretzky B, Eckstein W, Schorn RP (1995) Cu/Li Alloys: conditions for their application in fusion reactors—focussing on the interaction of segregation and sputtering phenomena. J Nucl Mater 224:50–70

Basile A, Hollenkamp AF, Bhatt AI, O'Mullane AP (2013) Extensive charge–discharge cycling of lithium metal electrodes achieved using ionic liquid electrolytes. Electrochem Commun 27:69–72. doi:10.1016/j.elecom.2012.10.030

Bates JB (1994) Protective lithium ion conducting ceramic coating for lithium metal anodes and associate method. USA Patent 5,314,765, 24 May 1994

Bates JB, Dudney NJ, Gruzalski GR, Zuhr RA, Choudhury A, Luck CF, Robertson JD (1993) Fabrication and Characterization of Amorphous Lithium Electrolyte Thin-Films and Rechargeable Thin-Film Batteries. J Power Sour 43(1–3):103–110. doi:10.1016/0378-7753 (93)80106-y

Belov DG, Yarmolenko OV, Peng A, Efimov ON (2006) Lithium surface protection by polyacetylene in situ polymerization. Syn Metals 156(9–10):745–751. doi:10.1016/j.synthmet. 2006.04.006

Besenhard JO (1977) The effect of I-anions on Li cycling in propylene carbonate. J Electroanal Chem 78:189–193

Besenhard JO (1978) Cycling behaviour and corrosion of Li–Al electrodes in organic electrolytes. J Electroanal Chem 1978(94):77–81

Besenhard JO, Eichinger G (1976) High energy density lithium cells. Part I. Electrolytes and anodes. J Electroanal Chem 68:1–18

Besenhard JO, Fritz HP, Wudy E, Dietz K, Meyer H (1985) Cycling of β-LiAl in organic electrolytes—effect of electrode contaminants and electrolyte additives. J Power Sour 14:193–200

Besenhard JO, Hess M, Komenda P (1990) Dimensionally stable Li-alloy electrodes for secondary batteries. Solid State Ionics 40(41):525–529

Besenhard JO, Wagner MW, Winter M, Jannakoudakis AD, Jannakoudakis PD, Theodoridou E (1993) Inorganic film-forming electrolyte additives improving the cycling behaviour of metallic lithium electrodes and the self-discharge of carbon—lithium electrodes. J Power Sour 44:413–420. doi:http://dx.doi.org/10.1016/0378-7753(93)80183-P

Best AS, Bhatt AI, Hollenkamp AF (2010) Ionic liquids with the Bis(fluorosulfonyl)imide anion: electrochemical properties and applications in battery technology. J Electrochem Soc 157 (8): A903–A911. doi:10.1149/1.3429886

Bhatt AI, Best AS, Huang J, Hollenkamp AF (2010) Application of the N-Propyl-N-methyl-pyrrolidinium Bis(fluorosulfonyl)imide RTIL containing lithium Bis (fluorosulfonyl)imide in ionic liquid based lithium batteries. J Electrochem Soc 157 (1): A66–A74. doi:10.1149/1.3257978

Bhatt AI, Kao P, Best AS, Hollenkamp AF (2012) Towards Li-air and Li-S batteries: understanding the morphological changes of lithium surfaces during cycling at a range of current densities in an ionic liquid electrolyte. ECS Trans 50(11):383–401

Bieker G, Winter M, Bieker P (2015) Electrochemical In situ investigations of SEI and dendrite formation on the lithium metal anode. Phys Chem Chem Phys 17 (14):8670–8679. doi:10. 1039/c4cp05865h

Bouchet R (2014) A stable lithium metal interface. Nat Nanotechnol 9:572–573

Brissot C, Rosso M, Chazalviel JN, Baudry P, Lascaud S (1998) In situ study of dendritic growth in lithium/PEO-salt/lithium cells. Electrochim Acta 43(10–11):1569–1574

Brissot C, Rosso M, Chazalviel J-N, Lascaud S (1999a) Dendritic growth mechanisms in lithium/polymer cells. J Power Sour 81–82:925–929

Brissot C, Rosso M, Chazalviel J-N, Lascaud S (1999b) In situ concentration cartography in the neighborhood of dendrites growing in lithium/polymer-electrolyte/lithium cells. J Electrochem Soc 146(12):4393–4400

Bronger W, Nacken B, Ploog K (1975) Zur Synthese und Struktur von Li_2Pt und LiPt. J Less-Common Metals 43:143–146

Bronger W, Klessen G, Müller P (1985) Zur Struktur von $LiPt_7$. J Less-Common Metals 109:L1–L2

Budi A, Basile A, Opletal G, Hollenkamp AF, Best AS, Rees RJ, Bhatt AI, O'Mullane AP, Russo SP (2012) Study of the initial stage of solid electrolyte interphase formation upon chemical reaction of lithium metal and N-methyl-N-propyl-pyrrolidinium-Bis(Fluorosulfonyl) Imide. J Phys Chem C 116 (37):19789–19797. doi:10.1021/jp304581g

Camurri C, Ortiz M, Carrasco C (2003) Hot consolidation of Cu–Li powder alloys: a first approach to characterization. Mater Charact 51 (2–3):171–176. doi:10.1016/j.matchar.2003.11.002

Cao C, Li Z-B, Wang X-L, Zhao X-B, Han W-Q (2014) Recent advances in inorganic solid electrolytes for lithium batteries. Front Energy Res 2:1–10. doi:10.3389/fenrg.2014.00025

Carpio RA, King LA (1981) Deposition and dissolution of lithium–aluminum alloy and aluminum from chloride-saturated LiCl–AlCl and NaCl–AlCl melts. J Electrochem Soc 128(7):1510–1517

Chang S-G, Lee HJ, Kang HY, Park S-M (2001) Characterization of surface films formed prior to bulk reduction of lithium in rigorously dried propylene carbonate solutions. Bull Korean Chem Soc 22(5):481–487

Chang HJ, Trease NM, Ilott AJ, Zeng D, Du L-S, Jerschow A, Grey CP (2015) Investigating Li microstructure formation on Li anodes for lithium batteries by in situ 6Li/7Li NMR and SEM. J Phys Chem C 119 (29):16443–16451. doi:10.1021/acs.jpcc.5b03396

Cheng H, Zhu C, Huang B, Lu M, Yang Y (2007) Synthesis and electrochemical characterization of PEO-based polymer electrolytes with room temperature ionic liquids. Electrochim Acta 52 (19):5789–5794. doi:10.1016/j.electacta.2007.02.062

Cheng L, Crumlin EJ, Chen W, Qiao R, Hou H, Franz Lux S, Zorba V, Russo R, Kostecki R, Liu Z, Persson K, Yang W, Cabana J, Richardson T, Chen G, Doeff M (2014) The origin of high electrolyte-electrode interfacial resistances in lithium cells containing garnet type solid electrolytes. Phys Chem Chem Phys 16 (34):18294–18300. doi:10.1039/c4cp02921f

Cheng L, Chen W, Kunz M, Persson K, Tamura N, Chen GY, Doeff M (2015a) Effect of surface microstructure on electrochemical performance of garnet solid electrolytes. Acs Appl Mater Inter 7 (3):2073–2081. doi:10.1021/am508111r

Choi N-S, Lee Y-M, Park J-H, Park J-K (2003) Interfacial enhancement between lithium electrode and polymer electrolytes. J Power Sour 119–121:610-616. doi:10.1016/s0378-7753(03)00305-7

Choi N-S, Lee YM, Cho KY, Ko D-H, Park J-K (2004a) Protective layer with Oligo(ethylene glycol) Borate anion receptor for lithium metal electrode stabilization. Electrochem Commun 6 (12):1238–1242. doi:10.1016/j.elecom.2004.09.023

Choi N-S, Lee YM, Seol W, Lee JA, Park J-K (2004b) Protective coating of lithium metal electrode for interfacial enhancement with gel polymer electrolyte. Solid State Ionics 172 (1–4):19–24. doi:10.1016/j.ssi.2004.05.008

Choi J-W, Kim J-K, Cheruvally G, Ahn J-H, Ahn H-J, Kim K-W (2007a) Rechargeable lithium/sulfur battery with suitable mixed liquid electrolytes. Electrochim Acta 52 (5):2075–2082. doi:10.1016/j.electacta.2006.08.016

Choi J, Cheruvally G, Kim Y, Kim J, Manuel J, Raghavan P, Ahn J, Kim K, Ahn H, Choi D (2007b) Poly(ethylene oxide)-based polymer electrolyte incorporating room-temperature ionic liquid for lithium batteries. Solid State Ionics 178 (19–20):1235–1241. doi:10.1016/j.ssi.2007.06.006

Choi J-W, Cheruvally G, Kim D-S, Ahn J-H, Kim K-W, Ahn H-J (2008) rechargeable lithium/sulfur battery with liquid electrolytes containing toluene as additive. J Power Sour 183 (1):441–445. doi:10.1016/j.jpowsour.2008.05.038

Choi N-S, Koo B, Yeon J-T, Lee KT, Kim D-W (2011) Effect of a novel amphipathic ionic liquid on lithium deposition in gel polymer electrolytes. Electrochim Acta 56 (21):7249–7255. doi:10.1016/j.electacta.2011.06.058

Choi SM, Kang IS, Sun Y-K, Song J-H, Chung S-M, Kim D-W (2013) Cycling characteristics of lithium metal batteries assembled with a surface modified lithium electrode. J Power Sour 244:363–368. doi:10.1016/j.jpowsour.2012.12.106

Christensen J, Albertus P, Sanchez-Carrera RS, Lohmann T, Kozinsky B, Liedtke R, Ahmed J, Kojic A (2012) A critical review of Li/Air batteries. J Electrochem Soc 159 (2):R1–R30. doi:10.1149/2.086202jes

Chung K-I, Lee J-D, Kim E-J, Kim W-S, Cho J-H, Choi Y-K (2003) Studies on the effects of coated Li$_2$CO$_3$ on lithium electrode. Microchem J 75 (2):71–77. doi:10.1016/s0026-265x(03)00026-2

Chung JH, Kim WS, Yoon WY, Min SW, Cho BW (2006) Electrolyte loss and dimensional change of the negative electrode in li powder secondary cell. J Power Sour 163 (1):191–195. doi:10.1016/j.jpowsour.2005.12.064

Dampier FW, Brummer SB (1977) The cycling behavior of the lithium electrode in LiAsF$_6$/methyl acetate solutions. Electrochim Acta 22:1339–1345

Dan P, Mengeritsky E, Aurbach D, Weissman I, Zinigrad E (1997) More details on the new LiMnO$_2$ rechargeable battery technology developed at Tadiran. J Power Sour 68:443–447

de Vries H, Jeong S, Passerini S (2015) Ternary polymer electrolytes incorporating pyrrolidinium-imide ionic liquids. RSC Adv 5(18):13598–13606. doi:10.1039/c4ra16070c

Débart A, Dupont L, Poizot P, Leriche JB, Tarascon JM (2001) A transmission electron microscopy study of the reactivity mechanism of tailor-made CuO particles toward lithium. J Electrochem Soc 148 (11):A1266–A1274. doi:10.1149/1.1409971

Devaux D, Harry KJ, Parkinson DY, Yuan R, Hallinan DT, MacDowell AA, Balsara NP (2015a) Failure mode of lithium metal batteries with a block copolymer electrolyte analyzed by X-ray microtomography. J Electrochem Soc 162 (7):A1301–A1309. doi:10.1149/2.0721507jes

Devaux D, Glé D, Phan TNT, Gigmes D, Giroud E, Deschamps M, Denoyel R, Bouchet R (2015b) Optimization of block copolymer electrolytes for lithium metal batteries. Chem Mater 27 (13):4682–4692. doi:10.1021/acs.chemmater.5b01273

Dey AN (1971) Electrochemical alloying of lithium in organic electrolytes. J Electrochem Soc 118 (10):1547–1549

Dey AN (1977) Lithium anode film and organic and inorganic electrolyte batteries. Thin Solid Films 43:131–171

Ding F, Xu W, Chen X, Zhang J, Engelhard MH, Zhang Y, Johnson BR, Crum JV, Blake TA, Liu X, Zhang JG (2013a) Effects of carbonate solvents and lithium salts on morphology and coulombic efficiency of lithium electrode. J Electrochem Soc 160 (10):A1894–A1901. doi:10.1149/2.100310jes

Ding F, Xu W, Graff GL, Zhang J, Sushko ML, Chen X, Shao Y, Engelhard MH, Nie Z, Xiao J, Liu X, Sushko PV, Liu J, Zhang J-G (2013b) Dendrite-free lithium deposition via self-healing electrostatic shield mechanism. J Am Chem Soc 135 (11):4450–4456. doi:10.1021/ja312241y

Dollé M, Sannier L, Beaudoin B, Trentin M, Tarascon J-M (2002) Live scanning electron microscope observations of dendritic growth in lithium/polymer cells. Electrochem Solid-State Lett 5 (12):A286–A289. doi:10.1149/1.1519970

Dominey LA, Goldman JL (1990) The improvement of rechargeable lithium battery electrolyte performance with additives. In: Proceedings of the 34th international power sources symposium 25 Jun 1990–28 Jun 1990, pp 84–86. doi:10.1109/IPSS.1990.145797

Dominey LA, Goldman JL, Koch VR, Shen D, Subbarao S, Huang CK, Halpert G, Deligiannis F (1991) Improved lithium/titanium disulfide cell cycling in either-based electrolytes with synergistic additives: part II. Proposed chemical pathways contributing to improved cycling. In: Abraham, KM, Salomon, M (eds) Proceedings of the symposium on primary and secondary lithium batteries. The Electrochem Soc Inc PV 91–3:293–301

Du F, Zhao N, Li Y, Chen C, Liu Z, Guo X (2015) All solid state lithium batteries based on lamellar garnet-type ceramic electrolytes. J Power Sour 300:24–28. doi:10.1016/j.jpowsour.2015.09.061

Dudley JT, Wilkinson DP, Thomas G, LeVae R, Woo S, Blom H, Horvath C, Juzkow MW, Denis B, Juric P, Aghakian P, Dahn JR (1991) Conductivity of electrolytes for rechargeable lithium batteries. J Power Sourc 35:59–82

Dudney NJ (2005) Solid-state thin-film rechargeable batteries. Mater Sci Eng B 116:245–249

Ebner WB, Lin HW (1987) Prototype rechargeable lithium batteries. US government report NSWC TR 86–108

Edström K, Herstedt M, Abraham DP (2006) A new look at the solid electrolyte interphase on graphite anodes in Li–Ion batteries. J Power Sour 153 (2):380–384. doi:10.1016/j.jpowsour.2005.05.062

Ein Ely Y, Aurbach D (1992) Identification of surface films formed on active metals and nonactive metal electrodes at low potentials in methyl formate solutions. Langmuir 8:1845–1850

Ein-Eli Y, Aurbach D (1996) The correlation between the cycling efficiency, surface chemistry and morphology of Li electrodes in electrolyte solutions based on methyl formate. J Power Sour 54:281–288

Ein-Eli Y, Thomas SR, Koch VR, Aurbach D, Markovsky B, Schechter A (1996) Ethylmethylcarbonate, a promising solvent for Li–Ion rechargeable batteries. J Electrochem Soc 143(12):L273–L277

Ein-Eli Y, McDevitt SF, Aurbach D, Markovsky B, Schechter A (1997) Methyl propyl carbonate: a promising single solvent for Li-ion battery electrolytes. J Electrochem Soc 144(7):L180–L184

Eweka E, Owen JR, Ritchie A (1997) Electrolytes and additives for high efficiency lithium cycling. J Power Sour 65:247–251

Fauteux D (1993) Lithium electrode in polymer electrolytes. Electrochim Acta 38(9):1199–1210

Fergus JW (2010) Ceramic and polymeric solid electrolytes for lithium-ion batteries. J Power Sour 195 (15):4554–4569. doi:10.1016/j.jpowsour.2010.01.076

Fernicola A, Croce F, Scrosati B, Watanabe T, Ohno H (2007) LiTFSI-BEPyTFSI as an improved ionic liquid electrolyte for rechargeable lithium batteries. J Power Sour 174 (1):342–348. doi:10.1016/j.jpowsour.2007.09.013

Ferrese A, Newman J (2014) Mechanical deformation of a lithium-metal anode due to a very stiff separator. J Electrochem Soc 161 (9):A1350–A1359. doi:10.1149/2.0911409jes

Fischer AK, Vissers DR (1983) Morphological studies on the Li–Al electrode in fused salt electrolytes. J Electrochem Soc 130(1):5–11

Frazer EJ (1981) Electrochemical formation of Lithium–Aluminium alloys in propylene carbonate electrolytes. J Electroanal Chem 121:329–339

Fu J (1997a) Lithium ion conductive glass-ceramics. USA Patent 5,702,995, 30 Dec 1997

Fu J (1997b) Superionic conductivity of glass-ceramics in the system Li_2O-Al_2O_3-TiO_2-P_2O_5. Solid State Ionics 96 (3–4):195–200. doi:10.1016/s0167-2738(97)00018-0

Fujieda T, Yamamoto N, Saito K, Ishibashi T, Honjo M, Koike S, Wakabayashi N, Higuchi S (1994) Surface of lithium electrodes prepared in Ar + CO_2 gas. J Power Sour 52:197–200. doi: http://dx.doi.org/10.1016/0378-7753(94)01961-4

Fujieda T, Koike S, Higuchi S (1998) Influence of water and other contaminants in electrolyte solutions on lithium electrodeposition. Mat Res Soc Symp Proc 496:463–468

Fung YS, Lai HC (1989) Cyclic chronopotentiometric studies of the LiAl anode in methyl acetate. J Appl Electrochem 19:239–246

Furuya R, Tachikawa N, Yoshii K, Katayama Y, Miura T (2015) Deposition and dissolution of lithium through lithium phosphorus oxynitride thin film in some ionic liquids. J Electrochem Soc 162 (9):H634–H637. doi:10.1149/2.0471509jes

Gachot G, Grugeon S, Armand M, Pilard S, Guenot P, Tarascon J-M, Laruelle S (2008) Deciphering the multi-step degradation mechanisms of carbonate-based electrolyte in Li batteries. J Power Sour 178 (1):409–421. doi:10.1016/j.jpowsour.2007.11.110

Gan H, Takeuchi ES (1996) Lithium electrodes with and without CO_2 treatment: electrochemical behavior and effect on high rate lithium battery performance. J Power Sour 62:45–50

Gao XP, Bao JL, Pan GL, Zhu HY, Huang PX, Wu F, Song DY (2004) Preparation and electrochemical performance of polycrystalline and single crystalline CuO nanorods as anode materials for Li Ion battery. J Phys Chem B 108:5547–5551

Garreau M, Thevenin J, Fekir M (1983) On the processes responsible for the degradation of the Aluminum–Lithium electrode used as anode material in lithium aprotic electrolyte batteries. J Power Sour 9:235–238

Gąsior W, Onderka B, Moser Z, Dębski A, Gancarz T (2009) Thermodynamic evaluation of Cu–Li phase diagram from EMF measurements and DTA study. Calphad 33 (1):215–220. doi:10.1016/j.calphad.2008.10.006

Gasparotto LH, Borisenko N, Bocchi N, El Abedin SZ, Endres F (2009) In situ STM investigation of the lithium underpotential deposition on Au(111) in the air- and water-stable ionic liquid 1-Butyl-1-methylpyrrolidinium Bis(trifluoromethylsulfonyl)amide. Phys Chem Chem Phys 11 (47):11140–11145. doi:10.1039/b916809e

Gauthier M, Fauteux D, Vassort G, Bélanger A, Duval M, Ricoux P, Chabagno J-M, Muller D, Rigaud P, Armand MB, Deroo D (1985a) Assessment of polymer-electrolyte batteries for EV and ambient temperature applications. J Electrochem Soc 132(6):1333–1340

Gauthier M, Fauteux D, Vassort G, Belanger A, Duval M, Ricoux P, Chabagno J-M, Muller D, Rigaud P, Armand MB, Deroo D (1985b) Behavior of polymer electrolyte batteries at 80–100 °C and near room temperature. J Power Sour 14:23–26

Geronov Y, Schwager F, Muller RH (1982) Electrochemical studies of the film formation on lithium in propylene carbonate solutions under open-circuit conditions. J Electrochem Soc 129 (7):1422–1429

Geronov Y, Zlatilova P, Moshtev RV (1984a) The secondary lithium-aluminum electrode at room temeperature. I. Cycling in LiClO4-propylene carbonate solutions. J Power Sour 12:145–153

Geronov Y, Zlatilova P, Staikov G (1984b) Electrochemical nucleation and growth of β-LiAl Alloy in aprotic electrolyte solutions. Electrochim Acta 29(4):551–555

Geronov Y, Zlatilova P, Staikov G (1984c) The secondary Lithium–Aluminum electrode at room temperature. II. Kinetics of the electrochemical formation of the Lithium–Aluminum Alloy. J Power Sour 12:155–165

Geronov Y, Zlatilova P, Puresheva B, Pasquali M, Pistoia G (1989) Behaviour of the Lithium electrode during cycling in nonaqueous solutions. J Power Sour 26:585–591

Girard GM, Hilder M, Zhu H, Nucciarone D, Whitbread K, Zavorine S, Moser M, Forsyth M, MacFarlane DR, Howlett PC (2015) Electrochemical and physicochemical properties of small phosphonium cation ionic liquid electrolytes with high lithium salt content. Phys Chem Chem Phys 17 (14):8706–8713. doi:10.1039/c5cp00205b

Gireaud L, Grugeon S, Laruelle S, Yrieix B, Tarascon JM (2006) Lithium metal stripping/plating mechanisms studies: a metallurgical approach. Electrochem Commun 8 (10):1639–1649. doi:10.1016/j.elecom.2006.07.037

Gofer Y, Ben-Zion M, Aurbach D (1992) Solutions of LiAsF6 in 1,3-Dioxolane for secondary lithium batteries. J Power Sour 39:163–178

Gofer Y, Barbour R, Luo Y, Tryk D, Scherson DA, Jayne J, Chottiner G (1995) Underpotential deposition of lithium on polycrystalline gold from a LiClO4/Poly(ethylene oxide) solid polymer electrolyte in ultrahigh vacuum. J Phys Chem 99:11739–11741

Goldman JL, Mank RM, Young JH, Koch VR (1980) Structure-reactivity relationships of methylated tetrahydrofurans with lithium. J Electrochem Soc 127(7):1461–1467

Goldman JL, Dominey LA, Koch VR (1989) The stabilization of LiAsF6/1,3-Dioxolane for use in rechargeable lithium batteries. J Power Sour 26:519–523

Goodman JKS, Kohl PA (2014) Effect of alkali and alkaline earth metal salts on suppression of lithium dendrites. J Electrochem Soc 161 (9):D418–D424. doi:10.1149/2.0301409jes

Gorodyskii AV, Sazhin SV, Danilin VV, Kuksenko SP (1989) Effect of sodium cation on lithium corrosion in aprotic media. J Power Sour 28:335–343

Grugeon S, Laruelle S, Herrera-Urbina R, Dupont L, Poizot P, Tarascon JM (2001) Particle size effects on the electrochemical performance of copper oxides toward lithium. J Electrochem Soc 148 (4):A285–A292. doi:10.1149/1.1353566

Guo J, Wen Z, Wu M, Jin J, Liu Y (2015) Vinylene carbonate–LiNO3: a hybrid additive in carbonic ester electrolytes for SEI modification on Li metal anode. Electrochem Commun 51:59–63. doi:10.1016/j.elecom.2014.12.008

Gurevitch I, Buonsanti R, Teran AA, Gludovatz B, Ritchie RO, Cabana J, Balsara NP (2013) Nanocomposites of titanium dioxide and polystyrene-Poly(ethylene oxide) block copolymer as solid-state electrolytes for lithium metal batteries. J Electrochem Soc 160 (9):A1611–A1617. doi:10.1149/2.117309jes

Hallinan DT, Mullin SA, Stone GM, Balsara NP (2013) Lithium metal stability in batteries with block copolymer electrolytes. J Electrochem Soc 160 (3):A464–A470. doi:10.1149/2.030303jes

Halpert G, Surampudi S, Shen D, Huang C-K, Narayanan S, Vamos E, Perrone D (1994) Status of the development of rechargeable lithium cells. J Power Sour 47:287–294

Hamon Y, Brousse T, Jousse F, Topart P, Buvat P, Schleich DM (2001) Aluminum negative electrode in lithium ion batteries. J Power Sour 97–98:185–187

Henderson WA (2006) Glyme–Lithium salt phase behavior. J Phys Chem B 110:13177–13183

Herbert EG, Tenhaeff WE, Dudney NJ, Pharr GM (2011) Mechanical characterization of LiPON films using nanoindentation. Thin Solid Films 520 (1):413–418. doi:10.1016/j.tsf.2011.07.068

Herlem G, Fahys B, Székely M, Sutter E, Mathieu C, Herlem M, Penneau J-F (1996) n-butylamine as solvent for lithium salt electrolytes. Structure and properties of concentrated solutions. Electrochim Acta 41(17):2753–2760

Herr R (1990) Organic electrolytes for lithium cells. Electrochim Acta 35(8):1257–1265

Hess S, Wohlfahrt-Mehrens M, Wachtler M (2015) Flammability of Li-Ion battery electrolytes: flash point and self-extinguishing time measurements. J Electrochem Soc 162 (2):A3084–A3097. doi:10.1149/2.0121502jes

Hirai T, Yoshimatsu I, Yamaki J-I (1994a) Influence of electrolyte on lithium cycling efficiency with pressurized electrode stack. J Electrochem Soc 141(3):611–614

Hirai T, Yoshimatsu I, Yamaki J-I (1994b) Effect of additives on lithium cycling efficiency. J Electrochem Soc 141(9):2300–2305

Honeywell I (1975) Lithium-organic electrolyte batteries for sensor and communications equipment. US Government report AD-A020 143

Hong S-T, Kim J-S, Lim S-J, Yoon WY (2004) Surface characterization of emulsified lithium powder electrode. Electrochim Acta 50 (2–3):535–539. doi:10.1016/j.electacta.2004.03.065

Howlett PC, MacFarlane DR, Hollenkamp AF (2004) High lithium metal cycling efficiency in a room-temperature ionic liquid. Electrochem Solid-State Lett 7 (5):A97–A101. doi:10.1149/1.1664051

Howlett PC, Brack N, Hollenkamp AF, Forsyth M, MacFarlane DR (2006) Characterization of the lithium surface in N-methyl-N-alkylpyrrolidinium Bis(trifluoromethanesulfonyl)amide room-temperature ionic liquid electrolytes. J Electrochem Soc 153 (3):A595–A606. doi:10.1149/1.2164726

Huang XH, Tu JP, Zhang B, Zhang CQ, Li Y, Yuan YF, Wu HM (2006) Electrochemical properties of NiO–Ni nanocomposite as anode material for lithium ion batteries. J Power Sour 161 (1):541–544. doi:10.1016/j.jpowsour.2006.03.039

Huang XH, Tu JP, Xia XH, Wang XL, Xiang JY, Zhang L, Zhou Y (2009) Morphology effect on the electrochemical performance of nio films as anodes for lithium ion batteries. J Power Sour 188 (2):588–591. doi:10.1016/j.jpowsour.2008.11.111

Huggins RA (1988) Polyphase alloys as rechargeable electrodes in advanced battery systems. J Power Sour 22:341–350

Huggins RA (1989a) Materials science principles related to alloys of potential use in rechargeable lithium cells. J Power Sour 26(1–2):109–120

Huggins RH (1989b) Materials science principles related to alloys of potential use in rechargeable lithium cells. J Power Sour 26:109–120

Huggins RA (1999a) Lithium alloy negative electrodes. J Power Sour 81–82:13–19

Huggins RA (1999b) Lithium alloy anodes in handbook of battery materials. Lithium Alloy anodes in handbook of battery materials (Ed: Besenhard, J O) Wiley-VCH:359–381

Huggins RA (2009) Lithium–Carbon alloys in advanced batteries-materials science aspects. Lithium–Carbon Alloys in Advanced Batteries-Materials Science Aspects Springer Science:127–149

Ichino T, Cahan BD, Scherson DA (1991) In situ attenuated total reflection fourier transform infrared spectroscopy studies of the polyethylene Oxide/LiClO$_4$-metallic lithium interface. J Electrochem Soc 138(11):L59–L61

Imanishi N, Hasegawa S, Zhang T, Hirano A, Takeda Y, Yamamoto O (2008) Lithium anode for lithium-air secondary batteries. J Power Sour 185 (2):1392–1397. doi:10.1016/j.jpowsour. 2008.07.080

Ishiguro K, Nakata Y, Matsui M, Uechi I, Takeda Y, Yamamoto O, Imanishi N (2013) Stability of Nb-doped cubic Li$_7$La$_3$Zr$_2$O$_{12}$ with Lithium Metal. J Electrochem Soc 160 (10):A1690–A1693. doi:10.1149/2.036310jes

Ishikawa M, Yoshitake S, Morita M, Matsuda Y (1994) In situ scanning vibrating electrode technique for the characterization of interface between lithium electrode and electrolytes containing additives. J Electrochem Soc 141(12):L159–L161

Ishikawa M, Morita M, Matsuda Y (1997) In situ scanning vibrating electrode technique for lithium metal anodes. J Power Sour 68:501–505

Ishikawa M, Kanemoto M, Morita M (1999a) Control of lithium metal anode cycleability by electrolyte temperature. J Power Sour 81–82:217–220

Ishikawa M, Machino S-I, Morita M (1999b) Electrochemical control of a Li metal anode interface: improvement of Li cyclability by inorganic additives compatible with electrolytes. J Electroanal Chem 473:279–284

Ishikawa M, Inoue K, Yoshimoto N, Morita M (2003) Cycleability enhancement of Li metal anode by primary charge with electrolyte additives. Electrochemistry 71(12):1046–1048

Ishikawa M, Kawasaki H, Yoshimoto N, Morita M (2005) Pretreatment of Li metal anode with electrolyte additive for enhancing Li cycleability. J Power Sour 146 (1–2):199–203. doi:10. 1016/j.jpowsour.2005.03.007

Ismail I, Noda A, Nishimoto A, Watanabe M (2001) XPS study of lithium surface after contact with lithium-salt doped polymer electrolytes. Electrochim Acta 46:1595–1603

Jang IC, Ida S, Ishihara T (2014) Surface coating layer on Li metal for increased cycle stability of Li-O$_2$ batteries. J Electrochem Soc 161 (5):A821–A826. doi:10.1149/2.087405jes

Jow TR, Liang CC (1982) Lithium–Aluminum electrodes at ambient temperatures. J Electrochem Soc 129(7):1429–1434

Jung YS, Oh DY, Nam YJ, Park KH (2015) Issues and challenges for bulk-type all-solid-state rechargeable lithium batteries using sulfide solid electrolytes. Isr J Chem 55 (5):472–485. doi:10.1002/ijch.201400112

Kamaya N, Homma K, Yamakawa Y, Hirayama M, Kanno R, Yonemura M, Kamiyama T, Kato Y, Hama S, Kawamoto K, Mitsui A (2011) A lithium superionic conductor. Nat Mater 10 (9):682–686. doi:10.1038/nmat3066

Kanamura K, Tamura H, Takehara Z-I (1992) XPS analysis of a lithium surface immersed in propylene carbonate solution containing various salts. J Electroanal Chem 333:127–142

Kanamura K, Shiraishi S, Takehara Z-I (1994a) Electrochemical deposition of uniform lithium on an Ni substrate in a nonaqueous electrolyte. J Electrochem Soc 141(9):L108–L110

Kanamura K, Shiraishi S, Tamura H, Takehara Z-I (1994b) X-Ray photoelectron spectroscopic analysis and scanning electron microscopic observation of the lithium surface immersed in nonaqueous solvents. J Electrochem Soc 141(9):2379–2385

Kanamura K, Shiraishi S, Takehara Z-I (1995a) Morphology control of lithium deposited in nonaqueous media. Chem Lett:209–210

Kanamura K, Tamura H, Shiraishi S, Takehara Z-I (1995b) XPS analysis for the lithium surface immersed in γ-butyrolactone containing various salts. Electrochim Acta 40(7):913–921

Kanamura K, Tamura H, Shiraishi S, Takehara Z-I (1995c) XPS analysis of lithium surfaces following immersion in various solvents containing LiBF$_4$. J Electrochem Soc 142(2):340–347

Kanamura K, Shiraishi S, Takeharo Z-I (1996) Electrochemical deposition of very smooth lithium using nonaqueous electrolytes containing HF. J Electrochem Soc 143(7):2187–2197

Kanamura K, Takezawa H, Shiraishi S, Takehara Z-I (1997) Chemical reaction of lithium surface during immersion in $LiClO_4$ or $LiPF_6$/DEC electrolyte. J Electrochem Soc 144(6):1900–1906

Kanamura K, Shiraishi S, Takehara Z-I (2000) Quartz crystal microbalance study of lithium deposition and dissolution in nonaqueous electrolyte with hydrofluoric acid. J Electrochem Soc 147(6):2070–2075

Kazyak E, Wood KN, Dasgupta NP (2015) Improved cycle life and stability of lithium metal anodes through ultrathin atomic layer deposition surface treatments. Chem Mater 27(18):6457–6462. doi:10.1021/acs.chemmater.5b02789

Khurana R, Schaefer JL, Archer LA, Coates GW (2014) Suppression of lithium dendrite growth using cross-linked polyethylene/poly(ethylene oxide) electrolytes: a new approach for practical lithium-metal polymer batteries. J Am Chem Soc 136(20):7395–7402. doi:10.1021/ja502133j

Kim JS, Yoon WY (2004a) Improvement in lithium cycling efficiency by using lithium powder anode. Electrochim Acta 50 (2–3):531–534. doi:10.1016/j.electacta.2003.12.071

Kim W-S, Yoon W-Y (2004b) Observation of dendritic growth on Li powder anode using optical cell. Electrochim Acta 50(2–3):541–545. doi:10.1016/j.electacta.2004.03.066

Kim SW, Ahn YJ, Yoon WY (1988) The surface morphology of Li metal electrode. Met Mater 6 (4):345–349. doi:10.1149/1.2095351

Kim J-S, Yoon W-Y, Kim B-K (2006) Morphological differences between lithium powder and lithium foil electrode during discharge/charge. J Power Sour 163 (1):258–263. doi:10.1016/j.jpowsour.2006.04.072

Kim G-T, Appetecchi GB, Alessandrini F, Passerini S (2007a) Solvent-free, PYR1ATFSI ionic liquid-based ternary polymer electrolyte systems. J Power Sour 171(2):861–869. doi:10.1016/j.jpowsour.2007.07.020

Kim JS, Yoon WY, Yi KY, Kim BK, Cho BW (2007b) The dissolution and deposition behavior in lithium powder electrode. J Power Sour 165 (2):620–624. doi:10.1016/j.jpowsour.2006.10.033

Kim GT, Appetecchi GB, Carewska M, Joost M, Balducci A, Winter M, Passerini S (2010a) UV cross-linked, lithium-conducting ternary polymer electrolytes containing ionic liquids. J Power Sour 195 (18):6130–6137. doi:10.1016/j.jpowsour.2009.10.079

Kim JS, Baek SH, Yoon WY (2010b) Electrochemical behavior of compacted lithium powder electrode in Li/V[sub 2]O[sub 5] rechargeable battery. J Electrochem Soc 157 (8):A984–A987. doi:10.1149/1.3457381

Kim TK, Chen W, Wang C (2011) Heat treatment effect of the Ni foam current collector in lithium ion batteries. J Power Sour 196 (20):8742–8746. doi:10.1016/j.jpowsour.2011.06.028

Kim GT, Jeong SS, Xue MZ, Balducci A, Winter M, Passerini S, Alessandrini F, Appetecchi GB (2012) Development of ionic liquid-based lithium battery prototypes. J Power Sour 199:239–246. doi:10.1016/j.jpowsour.2011.10.036

Kim TK, Li X, Wang C (2013a) Temperature dependent capacity contribution of thermally treated anode current collectors in lithium ion batteries. Appl Surf Sci 264:419–423. doi:10.1016/j.apsusc.2012.10.037

Kim S-H, Choi K-H, Cho S-J, Kil E-H, Lee S-Y (2013c) Mechanically compliant and lithium dendrite growth-suppressing composite polymer electrolytes for flexible lithium-ion batteries. J Mater Chem A 1 (16):4949–4955. doi:10.1039/c3ta10612h

Knauth P (2009) Inorganic solid Li ion conductors: an overview. Solid State Ionics 180(14–16):911–916. doi:10.1016/j.ssi.2009.03.022

Koch VR (1979) Reactions of tetrahydrofuran and lithium hexafluoroarsenate with lithium. J Electrochem Soc 126(2):181–187

Koch VR, Young JH (1978) The stability of the secondary lithium electrode in tetrahydrofuran-based electrolytes. J Electrochem Soc 125(9):1371–1377

Koch VR, Goldman JL, Mattos CJ, Mulvaney M (1982) Specular lithium deposits from lithium hexafluoroarsenate/diethyl ether electrolytes. J Electrochem Soc 129(1):1–4

Koike S, Fujieda T, Wakabayashi N, Higuchi S (1997) Electrochemical and quartz microbalance technique studies of anode material for secondary lithium batteries. J Power Sour 68:480–482

Kong SK, Kim BK, Yoon WY (2012) Electrochemical behavior of Li-powder anode in high Li capacity used. J Electrochem Soc 159(9):A1551–A1553. doi:10.1149/2.062209jes

Kwon CW, Cheon SE, Song JM, Kim HT, Kim KB, Shin CB, Kim SW (2001) Characteristics of a lithium-polymer battery based on a lithium powder anode. J Power Sour 93:145–150

Laman FC, Brandt K (1988) Effect of discharge current on cycle life of a rechargeable lithium battery. J Power Sour 24:195–206

Lambri OA, Peñaloza A, Morón Alcain AV, Ortiz M, Lucca FC (1996) Mechanical dynamical spectroscopy in Cu–Li alloys produced by electrodeposition. Mater Sci Engr A212:108–118

Lambri OA, Morón Alcain AV, Lambri GI, Peñaloza A, Ortiz M, Wörner CH, Bocanegra E (1999) Precipitation processes in a Cu-18 at% Li alloy produced by electrodeposition. Mater T JIM 40(1):72–77

Lambri OA, Pérez-Landazábal JI, Peñaloza A, Herrero O, Recarte V, Ortiz M, Wörner CH (2000) Study of the phases in a copper cathode during an electrodeposition process for obtaining Cu–Li Alloys. Mater Res Bull 35:1023–1033

Lambri OA, Pérez-Landazábal JI, Salvatierra LM, Recarte V, Bortolotto CE, Herrero O, Bolmaro RE, Peñaloza A, Wörner CH (2005) Obtaining of single phase Cu–Li alloy through an electrodeposition process. Mater Lett 59(2–3):349–354. doi:10.1016/j.matlet.2004.10.017

Lane GH, Best AS, MacFarlane DR, Forsyth M, Bayley PM, Hollenkamp AF (2010a) the electrochemistry of lithium in ionic liquid/organic diluent mixtures. Electrochim Acta 55 (28):8947–8952. doi:10.1016/j.electacta.2010.08.023

Lane GH, Bayley PM, Clare BR, Best AS, MacFarlane DR, Forsyth M, Hollenkamp AF (2010b) Ionic liquid electrolyte for lithium metal batteries: physical, electrochemical, and interfacial studies of N-Methyl-N-butylmorpholinium Bis(fluorosulfonyl)imide. J Phys Chem C 114:21775–21785

Lee H, Cho J-J, Kim J, Kim H-J (2005) Comparison of voltammetric responses over the cathodic region in LiPF6 and LiBETI with and without HF. J Electrochem Soc 152(6):A1193–A1198. doi:10.1149/1.1914748

Lee Y-G, Kyhm K, Choi N-S, Ryu KS (2006) Submicroporous/microporous and compatible/incompatible multi-functional dual-layer polymer electrolytes and their interfacial characteristics with lithium metal anode. J Power Sour 163(1):264–268. doi:10.1016/j.jpowsour.2006.05.008

Lee Y-S, Lee JH, Choi J-A, Yoon WY, Kim D-W (2013) Cycling characteristics of lithium powder polymer batteries assembled with composite gel polymer electrolytes and lithium powder anode. Adv Funct Mater 23(8):1019–1027. doi:10.1002/adfm.201200692

Lee H, Lee DJ, Kim Y-J, Park J-K, Kim H-T (2015) A simple composite protective layer coating that enhances the cycling stability of lithium metal batteries. J Power Sour 284:103–108. doi:10.1016/j.jpowsour.2015.03.004

Li J, Pons S, Smith JJ (1986) Far-infrared spectroscopy of the electrode-solution with surface states of a gold electrode. Langmuir 2:297–301

Li L-F, Totir DA, Chottiner GS, Scherson DA (1998a) Electrochemical reactivity of carbon monoxide and sulfur adsorbed on Ni(111) and Ni(110) in a lithium-based solid polymer electrolyte in ultrahigh vacuum. J Phys Chem B 102:8013–8016

Li L-F, Totir DA, Gofer Y, Chottiner GS, Scherson DA (1998b) The electrochemistry of nickel in a lithium-based solid polymer electrolyte in ultrahigh vacuum environments. Electrochim Acta 44:949–955

Li L-F, Luo Y, Totir GG, Totir DA, Chottiner GS, Scherson DA (1999) Underpotential deposition of lithium on aluminum in ultrahigh-vacuum environments. J Phys Chem B 103:164–168

Li X, Dhanabalan A, Bechtold K, Wang C (2010) Binder-free porous core-shell structured Ni/NiO configuration for application of high performance lithium ion batteries. Electrochem Commun 12(9):1222–1225. doi:10.1016/j.elecom.2010.06.024

Li L, Zhao X, Fu Y, Manthiram A (2012a) Polyprotic acid catholyte for high capacity dual-electrolyte Li–air batteries. Phys Chem Chem Phys 14:12737–12740

Li Y, Huang K, Xing Y (2012b) A hybrid Li-air battery with buckypaper air cathode and sulfuric acid electrolyte. Electrochim Acta 81:20–24

Li W, Yao H, Yan K, Zheng G, Liang Z, Chiang YM, Cui Y (2015) The synergetic effect of lithium polysulfide and lithium nitrate to prevent lithium dendrite growth. Nat Commun 6:7436. doi:10.1038/ncomms8436

Lisowska-Oleksiak A (1999) The interface between lithium and poly(ethylene-oxide). Solid State Ionics 119:205–209

Liu S, Imanishi N, Zhang T, Hirano A, Takeda Y, Yamamoto O, Yang J (2010a) Effect of nano-silica filler in polymer electrolyte on Li dendrite formation in Li/Poly(ethylene oxide)–Li (CF3SO2)2 N/Li. J Power Sour 195(19):6847–6853. doi:10.1016/j.jpowsour.2010.04.027

Liu S, Imanishi N, Zhang T, Hirano A, Takeda Y, Yamamoto O, Yang J (2010b) Lithium dendrite formation in Li/Poly(ethylene oxide)–Lithium Bis(trifluoromethanesulfonyl)imide and N-methyl-N-propylpiperidinium Bis(trifluoromethanesulfonyl)imide/Li cells. J Electrochem Soc 157(10):A1092–A1098. doi:10.1149/1.3473790

Liu H, Wang G, Liu J, Qiao S, Ahn H (2011a) Highly ordered mesoporous NiO anode material for lithium ion batteries with an excellent electrochemical performance. J Mater Chem 21 (9):3046–3052. doi:10.1039/c0jm03132a

Liu S, Wang H, Imanishi N, Zhang T, Hirano A, Takeda Y, Yamamoto O, Yang J (2011b) Effect of Co-doping nano-silica filler and N-methyl-N-propylpiperidinium Bis(trifluoromethanesul-fonyl)imide into polymer electrolyte on Li dendrite formation in Li/Poly(ethylene oxide)-Li (CF3SO2)2 N/Li. J Power Sour 196(18):7681–7686. doi:10.1016/j.jpowsour.2011.04.001

Liu Z, Fu W, Payzant EA, Yu X, Wu Z, Dudney NJ, Kiggans J, Hong K, Rondinone AJ, Liang C (2013) Anomalous high ionic conductivity of nanoporous β-Li3PS4. J Am Chem Soc 135 (3):975–978. doi:10.1021/ja3110895

Liu C, Ma X, Xu F, Zheng L, Zhang H, Feng W, Huang X, Armand M, Nie J, Chen H, Zhou Z (2014) Ionic liquid electrolyte of lithium Bis(fluorosulfonyl) imide/N-methyl-N-propylpiperidinium Bis(fluorosulfonyl)imide for Li/natural graphite cells: effect of concentration of lithium salt on the physicochemical and electrochemical properties. Electrochim Acta 149:370–385. doi:10.1016/j.electacta.2014.10.048

Rendek Jr. LJ, Chottiner GS, Scherson DA (2003) Reactivity of metallic lithium toward γ-butyrolactone, propylene carbonate, and dioxalane. J Electrochem Soc 150 (3):A326–A329. doi:10.1149/1.1543949

Lu Y, Tu Z, Archer LA (2014) Stable lithium electrodeposition in liquid and nanoporous solid electrolytes. Nat Mater 13(10):961–969. doi:10.1038/nmat4041

Lu Y, Tu Z, Shu J, Archer LA (2015) Stable lithium electrodeposition in salt-reinforced electrolytes. J Power Sour 279:413–418. doi:10.1016/j.jpowsour.2015.01.030

Lv D, Shao Y, Lozano T, Bennett WD, Graff GL, Polzin B, Zhang J-G, Engelhard MH, Saenz NT, Henderson WA, Bhattacharya P, Liu J, Xiao J (2015) Failure mechanism for fast-charged lithium metal batteries with liquid electrolytes. Adv Energy Mater 5(3):1400993. doi:10.1002/aenm.201400993

Malik Y, Aurbach D, Dan P, Meitav A (1990) The electrochemical behaviour of 2-methyltetrahydrofuran solutions. J Electroanal Chem 282:73–105

Marcinek M, Syzdek J, Marczewski M, Piszcz M, Niedzicki L, Kalita M, Plewa-Marczewska A, Bitner A, Wieczorek P, Trzeciak T, Kasprzyk M, Łężak P, Zukowska Z, Zalewska A, Wieczorek W (2015) Electrolytes for Li-Ion transport—review. Solid State Ionics 276:107–126. doi:10.1016/j.ssi.2015.02.006

Marlier JF, Frey TG, Mallory JA, Cleland WW (2005) Multiple isotope effect study of the acid-catalyzed hydrolysis of methyl formate. J Org Chem 70:1737–1744

Matsuda Y (1993) Behavior of lithium/electrolyte interface in organic solutions. J Power Sour 43–44:1–7

Matsuda Y, Morita M (1989) Organic additives for the electrolytes of rechargeable lithium batteries. J Power Sour 26:579–583. doi:http://dx.doi.org/10.1016/0378-7753(89)80182-X

Matsuda Y, Sekiya M (1999) Effect of organic additives in electrolyte solutions on lithium electrode behavior. J Power Sour 81–82:759–761. doi:http://dx.doi.org/10.1016/S0378-7753 (99)00239-6

Matsuda Y, Morita M, Nigo H (1991) In: Abraham KM, Salomon M (eds) Primary and secondary lithium batteries. The electrochemical society proceedings series, Pennington 91–3:272

Matsuda Y, Ishikawa M, Yoshitake S, Morita M (1995) Characterization of the lithium-organic electrolyte interface containing inorganic and organic additives by in situ techniques. J Power Sour 54:301–305

Matsuda Y, Monta M, Ishikawa M (1997) Electrolyte solutions for anodes in rechargeable lithium batteries. J Power Sourc 68:30–36

Matsui T, Takeyama K (1995) Lithium deposit morphology from polymer electrolytes. Electrochim Acta 40(13–14):2165–2169

Matsumoto H, Sakaebe H, Tatsumi K, Kikuta M, Ishiko E, Kono M (2006) Fast Cycling of Li/LiCoO$_2$ cell with low-viscosity ionic liquids based on Bis(fluorosulfonyl)imide [FSI]. J Power Sour 160 (2):1308–1313. doi:10.1016/j.jpowsour.2006.02.018

Mayers MZ, Kaminski JW, Miller TF (2012) Suppression of dendrite formation via pulse charging in rechargeable lithium metal batteries. J Phys Chem C 116 (50):26214–26221. doi:10.1021/jp309321w

McKinnon WR, Dahn JR (1985) How to reduce the cointercalation of propylene carbonate in Li $_x$ ZrS$_2$ and other layered compounds. J Electrochem Soc 132(2):364–366

McOwen DW, Seo DM, Borodin O, Vatamanu J, Boyled PD, Henderson WA (2014) Concentrated electrolytes: decrypting electrolyte properties and reassessing Al corrosion mechanisms. Energy Environ Sci 7:416

Mengeritsky E, Dan P, Weissman I, Zaban A, Aurbach D (1996a) Safety and performance of Tadiran TLR-7103 rechargeable batteries. J Electrochem Soc 143 (7):2110–2116. doi:10.1149/1.1836967

Mengeritsky E, Dan P, Weissman I, Zaban A, Aurbach D (1996b) Safety and performance of Tadiran TLR-7103 rechargeable batteries. J Electrochem Soc 143(7):2110–2116

Minami T, Hayashi A, Tatsumisago M (2006) Recent progress of glass and glass-ceramics as solid electrolytes for lithium secondary batteries. Solid State Ionics 177 (26–32):2715–2720. doi:10.1016/j.ssi.2006.07.017

Mo Y, Gofer Y, Hwang E, Wang Z-G, Scherson DA (1996) Simultaneous microgravimetric and optical reflectivity studies of lithium underpotential deposition on Au(111) from propylene carbonate electrolytes. J Electroanal Chem 409:87–93

Mogi R, Inaba M, Abe T, Ogumi Z (2001) In situ atomic force microscopy observation of lithium deposition at an elevated temperature. J Power Sour 97–98:265–268

Mogi R, Inaba M, Iriyama Y, Abe T, Ogumi Z (2002a) In situ atomic force microscopy study on lithium deposition on nickel substrates at elevated temperatures. J Electrochem Soc 149 (4):A385–A390. doi:10.1149/1.1454138

Mogi R, Inaba M, Iriyama Y, Abe T, Ogumi Z (2002b) Surface film formation on nickel electrodes in a propylene carbonate solution at elevated temperatures. J Power Sour 108:163–173

Mogi R, Inaba M, Jeong S-K, Iriyama Y, Abe T, Ogumi Z (2002c) Effects of some organic additives on lithium deposition in propylene carbonate. J Electrochem Soc 149 (12):A1578–A1583. doi:10.1149/1.1516770

Momma T, Matsumoto Y, Osaka T (1995) Effect of CO$_2$ on the cycleability of lithium metal anode. Mat Res Soc Symp Proc 393:223–228

Momma T, Nara H, Yamagami S, Tatsumi C, Osaka T (2011) Effect of the atmosphere on chemical composition and electrochemical properties of solid electrolyte interface on electrodeposited Li metal. J Power Sour 196 (15):6483–6487. doi:10.1016/j.jpowsour.2011.03.095

Monroe C, Newman J (2005) the impact of elastic deformation on deposition kinetics at lithium/polymer interfaces. J Electrochem Soc 152 (2):A396–A404. doi:10.1149/1.1850854

Morales J, Trócoli R, Franger S, Santos-Peña J (2010) Cycling-induced stress in lithium ion negative electrodes: LiAl/LiFePO$_4$ and Li$_4$Ti$_5$O$_{12}$/LiFePO$_4$ cells. Electrochim Acta 55 (9):3075–3082. doi:10.1016/j.electacta.2009.12.104

Morigaki K-I, Ohta A (1998) Analysis of the surface of lithium in organic electrolyte by atomic force microscopy, fourier transform infrared spectroscopy and scanning auger electron microscopy. J Power Sour 76:159–166

Morita M, Aoki S, Matsuda Y (1992) AC impedance behavior of lithium electrode in organic electrolyte solutions containing additives. Electrochim Acta 37(1):119–123

Motoyama M, Ejiri M, Iriyama Y (2015) Modeling the nucleation and growth of Li at metal current collector/LiPON interfaces. J Electrochem Soc 162 (13):A7067–A7071. doi:10.1149/2.0051513jes

Murugan R, Thangadurai V, Weppner W (2007) Fast lithium ion conduction in garnet-type Li7La3Zr2O12. Angew Chem-Int Ed 46 (41):7778–7781. doi:10.1002/anie.200701144

Myung S-T, Yashiro H (2014) Electrochemical stability of aluminum current collector in alkyl carbonate electrolytes containing lithium Bis(pentafluoroethylsulfonyl)imide for lithium-ion batteries. J Power Sour 271:167–173. doi:10.1016/j.jpowsour.2014.07.097

Myung S-T, Natsui H, Sun Y-K, Yashiro H (2010) Electrochemical behavior of Al in a non-aqueous alkyl carbonate solution containing LiBOB salt. J Power Sour 195 (24):8297–8301. doi:10.1016/j.jpowsour.2010.07.027

Nagao M, Kitaura H, Hayashi A, Tatsumisago M (2009) J Power Sour 189:672–675

Nagao M, Hayashi A, Tatsumisago M, Kanetsuku T, Tsuda T, Kuwabata S (2013a) In situ SEM study of a lithium deposition and dissolution mechanism in a bulk-type solid-state cell with a Li2S-P2S5 solid electrolyte. Phys Chem Chem Phys 15 (42):18600–18606. doi:10.1039/c3cp51059j

Nanjundiah C, Goldman JL, Dominey LA, Koch VR (1988) Electrochemical stability of LiMF6 (M = P, As, Sb) in tetrahydrofuran and sulfolane. J Electrochem Soc 135(12):2914–2917

Nazri G, Muller RH (1985a) In situ X-ray diffraction of surface layers on lithium in nonaqueous electrolyte. J Electrochem Soc 132(6):1385–1387

Nazri G, Muller RH (1985b) Effect of residual water in propylene carbonate on films formed on lithium. J Electrochem Soc 132(9):2054–2058

Needham SA, Wang GX, Liu HK (2006) Synthesis of NiO nanotubes for use as negative electrodes in lithium ion batteries. J Power Sour 159 (1):254–257. doi:10.1016/j.jpowsour.2006.04.025

Nemori H, Matsuda Y, Mitsuoka S, Matsui M, Yamamoto O, Takeda Y, Imanishi N (2015) Stability of Garnet-type solid electrolyte LixLa3A2-yByO12 (A = Nb or Ta, B = Sc or Zr). Solid State Ionics 282:7–12. doi:10.1016/j.ssi.2015.09.015

Neudecker BJ, Dudney NJ, Bates JB (2000) "Lithium-Free" thin-film battery with in situ plated Li anode. J Electrochem Soc 147(2):517–523

Neuhold S, Vaughey JT, Grogger C, López CM (2014) Enhancement in cycle life of metallic lithium electrodes protected with Fp-Silanes. J Power Sour 254:241–248. doi:10.1016/j.jpowsour.2013.12.057

Newman GH, Francis RW, Gaines LH, Rao BML (1980) Hazard Investigations of LiClO4/Dioxolane electrolyte. J Electrochem Soc 127(9):2025–2027

Niitani T, Shimada M, Kawamura K, Kanamura K (2005a) Characteristics of new-type solid polymer electrolyte controlling nano-structure. J Power Sour 146 (1–2):386–390. doi:10.1016/j.jpowsour.2005.03.102

Niitani T, Shimada M, Kawamura K, Dokko K, Rho Y-H, Kanamura K (2005b) Synthesis of Li + Ion conductive PEO-PSt block copolymer electrolyte with microphase separation structure. Electrochem Solid-State Lett 8 (8):A385–A388. doi:10.1149/1.1940491

Niitani T, Amaike M, Nakano H, Dokko K, Kanamura K (2009) Star-shaped polymer electrolyte with microphase separation structure for all-solid-state lithium batteries. J Electrochem Soc 156 (7):A577–A583. doi:10.1149/1.3129245

Nishida T, Nishikawa K, Rosso M, Fukunaka Y (2013) Optical observation of Li dendrite growth in ionic liquid. Electrochim Acta 100:333–341. doi:10.1016/j.electacta.2012.12.131

Nishikawa K, Fukunaka Y, Sakka T, Ogata YH, Selman JR (2007) Measurement of concentration profiles during electrodeposition of Li metal from LiPF6-PC electrolyte solution. J Electrochem Soc 154 (10):A943–A948. doi:10.1149/1.2767404

Nishikawa K, Mori T, Nishida T, Fukunaka Y, Rosso M, Homma T (2010) In situ observation of dendrite growth of electrodeposited li metal. J Electrochem Soc 157 (11):A1212–A1217. doi:10.1149/1.3486468

Nishikawa K, Mori T, Nishida T, Fukunaka Y, Rosso M (2011) Li dendrite growth and Li + Ionic mass transfer phenomenon. J Electroanal Chem 661 (1):84–89. doi:10.1016/j.jelechem.2011.06.035

Nishio Y, Kitaura H, Hayashi A, Tatsumisago M (2009) All-solid-state lithium secondary batteries using nanocomposites of NiS electrode/Li_2S-P_2S_5 electrolyte prepared mechanochemical reaction. J Power Sour 189:629–632

Ogumi Z, Wang H (2009). In: Yoshio, M, Brodd RJ, Kozawa A (eds) Carbon anode materials in Lithium-Ion batteries: science & technologies. Springer Science, pp 49–73

Ohta N, Takada K, Zhang L, Ma R, Osada M, Sasaki T (2006) Enhancement of the high-rate capability of solid-state lithium batteries by nanoscale interfacial modification. Adv Mater 18 (17):2226–2229. doi:10.1002/adma.200502604

Ohta S, Kobayashi T, Seki J, Asaoka T (2012) Electrochemical performance of an all-solid-state lithium ion battery with garnet-type oxide electrolyte. J Power Sour 202:332–335. doi:10.1016/j.jpowsour.2011.10.064

Ohta S, Komagata S, Seki J, Saeki T, Morishita S, Asaoka T (2013) All-solid-state lithium ion battery using garnet-type oxide and Li3BO3 solid electrolytes fabricated by screen-printing. J Power Sour 238:53–56. doi:10.1016/j.jpowsour.2013.02.073

Okamoto H (1989) The C–Li (Carbon–Lithium) system. Bull Alloy Phase Diag 10(1):69–72

Okamoto H (1993) Li–Pb (Lithium–Lead). J Phase Equilib 14(6):770

Okamoto H (2012a) Li–Zn (Lithium–Zinc). J Phase Equilib Diff 33 (4):345–345. doi:10.1007/s11669-012-0052-x

Okamoto H (2012b) Al–Li (Aluminum–Lithium). J Phase Equilib Diff 33 (6):500–501. doi:10.1007/s11669-012-0119-8

Orsini F, Dollé M, Tarascon JM (2000) Impedance study of the Li/electrolyte interface upon cycling. Solid State Ionics 135:213–221

Osaka T, Momma T, Matsumoto Y, Uchida Y (1997a) Surface characterization of electrodeposited lithium anode with enhanced cycleability obtained by CO_2 addition. J Electrochem Soc 144 (5):1709–1713

Osaka T, Momma T, Matsumoto Y, Uchida Y (1997b) Effect of carbon dioxide on lithium anode cycleability with various substrates. J Power Sour 68:497–500

Osaka T, Kitahara M, Uchida Y, Momma T, Nishimura K (1999) Improved morphology of plated lithium in Poly(vinylidene fluoride) based electrolyte. J Power Sour 81–82:734–738

Ota M, Izuo S, Nishikawa K, Fukunaka Y, Kusaka E, Ishii R, Selman JR (2003) Measurement of concentration boundary layer thickness development during lithium electrodeposition onto a lithium metal cathode in propylene carbonate. J Electroanal Chem 559:175–183. doi:10.1016/j.jelechem.2003.08.020

Ota H, Wang X, Yasukawa E (2004a) Characterization of lithium electrode in lithium imides/ethylene carbonate, and cyclic ether electrolytes. I. surface morphology and lithium cycling efficiency. J Electrochem Soc 151 (3):A427–A436. doi:10.1149/1.1644136

Ota H, Shima K, Ue M, Yamaki J-I (2004b) Effect of vinylene carbonate as additive to electrolyte for lithium metal anode. Electrochim Acta 49 (4):565–572. doi:10.1016/j.electacta.2003.09.010

Ota H, Sakata Y, Wang X, Sasahara J, Yasukawa E (2004c) Characterization of lithium electrode in lithium imides/ethylene carbonate and cyclic ether electrolytes. II. Surface chemistry. J Electrochem Soc 151 (3):A437–A446. doi:10.1149/1.1644137

Ota H, Sakata Y, Otake Y, Shima K, Ue M, Yamaki J (2004d) Structural and functional analysis of surface film on Li anode in vinylene carbonate-containing electrolyte. J Electrochem Soc 151 (11):A1778–A1788. doi:10.1149/1.1798411

Paddon CA, Compton RG (2007) Underpotential deposition of lithium on platinum single crystal electrodes in tetrahydrofuran. J Phys Chem C 111:9016–9018

Pappenfus TM, Henderson WA, Owens BB, Mann KR, Smyrl WH (2004) Complexes of lithium imide salts with tetraglyme and their polyelectrolyte composite materials. J Electrochem Soc 151 (2):A209–A215. doi:10.1149/1.1635384

Park MS, Yoon WY (2003) Characteristics of a Li/MnO$_2$ battery using a lithium powder anode at high-rate discharge. J Power Sour 114 (2):237–243. doi:10.1016/s0378-7753(02)00581-5

Park SH, Winnick J, Kohl PA (2002) Investigation of the lithium couple on Pt, Al, and Hg electrodes in lithium imide-ethyl methyl sulfone. J Electrochem Soc 149 (9):A1196–A1200. doi:10.1149/1.1497979

Pastorello S (1930) Thermal analysis of the system: lithium-copper. Gazz Chim Ital 60:988–992

Peled E (1979) The electrochemical behavior of alkali and alkaline earth metals in nonaqueous battery systems—The solid electrolyte interphase model. J Electrochem Soc 126(12):2047–2051

Peled E (1983) Film forming reaction at the lithium/electrolyte interface. J Power Sour 9:253–266

Peled E, Sternberg Y, Gorenshtein A, Lavi Y (1989) Lithium-sulfur battery: evaluation of dioxolane-based electrolytes. J Electrochem Soc 136(6):1621–1625

Peled E, Golodnitsky D, Ardel G (1997) Advanced model for solid electrolyte interphase electrodes in liquid and polymer electrolytes. J Electrochem Soc 144(8):L208–L210

Pelton AD (1986a) The Au–Li (Gold–Lithium) system. Bull Alloy Phase Diag 7(3):228–231

Pelton AD (1986b) The Ag–Li (Silver–Lithium) system. Bull Alloy Phase Diag 7(3):223–228

Pelton AD (1986c) The Cu–Li (Copper–Lithium) system. Bull Alloy Phase Diag 7(2):142–144

Pelton AD (1988) The Cd–Li (Cadmium–Lithium) system. Bull Alloy Phase sDiag 9(1):36–41

Pelton AD (1991) The Li–Zn (Lithium–Zinc) system. J Phase Equilib 12(1):42–45

Peñaloza À, Ortíz M, Wörner CH (1995) An electrodeposition method to obtain Cu–Li alloys. J Mater Sci Lett 14:511–513

Pérez-Landazábal JI, Lambri OA, Peñaloza A, Recarte V, Campo J, Salvatierra LM, Herrero O, Ortiz M, Milani LM, Wörner CH (2002) Effect of the oxygen in the evolution of the microstructure in a Cu–18 at% Li Alloy. Mater Lett 56:709–715

Pletcher D, Rohan JF, Ritchie AG (1994) Microelectrode studies of the lithium/propylene carbonate system—Part I. Electrode reactions at potentials positive to lithium deposition. Electrochim Acta 39(10):1369–1376

Plichta E, Salomon M, Slane S, Uchiyama M (1987) Conductance of 1:1 electrolytes in methyl formate. J Soln Chem 16(3):225–235

Plichta E, Slane S, Uchiyama M, Salomon M, Chua D, Ebner WB, Lin HW (1989) An improved Li/LixCoO$_2$ rechargeable cell. J Electrochem Soc 136(7):1865–1869

Predel B (1997a) Li–Rb (Lithium–Rubidium). Li–Rb (Lithium–Rubidium) in phase equilibria, crystallographic and thermodynamic data of binary alloys · Li–Mg—Nd–Zr (Landolt-Börnstein —Group IV Physical Chemistry Volume 5H) (Ed Madelung, O)

Predel B (1997b) Li–Pd (Lithium–Palladium). Li–Pd (Lithium–Palladium) in phase equilibria, crystallographic and thermodynamic data of binary alloys Li–Mg—Nd–Zr (Landolt-Börnstein —Group IV physical chemistry volume 5H) (Ed Madelung, O)

Predel B (1997c) Li–Mo (Lithium–Molybdenum). Li–Mo (Lithium–Molybdenum) in phase equilibria, crystallographic and thermodynamic data of binary alloys · Li–Mg—Nd–Zr (Landolt-Börnstein—Group IV physical chemistry volume 5H) (Ed Madelung, O). doi:10. 1007/10522884_1907

Predel B (1997d) Li–Ni (Lithium–Nickel). Li–Ni (Lithium–Nickel) in phase equilibria, crystallographic and thermodynamic data of binary alloys · Li–Mg—Nd–Zr (Landolt-Börnstein—Group IV Physical Chemistry Volume 5H) (Ed Madelung, O)

Predel B (1997e) Li–Pb (Lithium–Lead). Li–Pb (Lithium–Lead) in phase equilibria, crystallographic and thermodynamic data of binary alloys · Li–Mg—Nd–Zr (Landolt-Börnstein— Group IV physical chemistry volume 5H) (Ed Madelung, O)

Predel B (1997f) Li–Ti (Lithium–Titanium). Li–Ti (Lithium–Titanium) in phase equilibria, crystallographic and thermodynamic data of binary alloys · Li–Mg—Nd–Zr (Landolt-Börnstein —Group IV physical chemistry volume 5H) (Ed Madelung, O)

Predel B (1997g) Li–Mn (Lithium–Manganese). Li–Mn (Lithium–Manganese) in phase equilibria, crystallographic and thermodynamic data of binary alloys · Li–Mg—Nd–Zr (Landolt-Börnstein —Group IV physical chemistry volume 5H) (Ed Madelung, O)

Predel B (1997h) Li–Mo (Lithium–Molybdenum). Li–Mo (Lithium–Molybdenum) in phase equilibria, crystallographic and thermodynamic data of binary alloys · Li–Mg—Nd–Zr (Landolt-Börnstein—Group IV physical chemistry volume 5H) (Ed Madelung, O)

Qian J, Xu W, Bhattacharya P, Engelhard M, Henderson WA, Zhang Y, Zhang J-G (2015a) Dendrite-Free Li deposition using trace-amounts of water as an electrolyte additive. Nano Energy 15:135–144. doi:10.1016/j.nanoen.2015.04.009

Qian J, Henderson WA, Xu W, Bhattacharya P, Engelhard M, Borodin O, Zhang JG (2015b) High rate and stable cycling of lithium metal anode. Nat Commun 6:6362. doi:10.1038/ncomms7362

Rao BML, Francis RW, Christopher HA (1977) Lithium–Aluminum electrode. J Electrochem Soc 124(10):1490–1492

Rauh RD (1975) Some observations on the attack of esters by lithium. US Government report AD/A-006 746

Rauh RD, Brummer SB (1977a) The effect of additives on lithium cycling in propylene carbonate. Electrochim Acta 22:75–83

Rauh RD, Brummer SB (1977b) The effect of additives on lithium cycling in methyl acetate. Electrochim Acta 22:85–91

Ren Y, Shen Y, Lin Y, Nan C-W (2015a) Direct observation of lithium dendrites inside garnet-type Lithium–Ion solid electrolyte. Electrochem Commun 57:27–30. doi:10.1016/j.elecom.2015.05.001

Ren Y, Chen K, Chen R, Liu T, Zhang YH, Nan CW (2015b) Oxide electrolytes for lithium batteries. J Am Ceramic Soc:1–21. doi:10.1111/jace.13844

Roberts M, Younesi R, Richardson W, Liu J, Zhu J, Edstrom K, Gustafsson T (2014) Increased cycling efficiency of lithium anodes in dimethyl sulfoxide electrolytes for use in Li-O$_2$ batteries. ECS Electrochem Lett 3 (6):A62–A65. doi:10.1149/2.007406eel

Ryou M-H, Lee YM, Lee Y, Winter M, Bieker P (2015) Mechanical surface modification of lithium metal: towards improved Li metal anode performance by directed Li plating. Adv Funct Mater 25(6):834–841. doi:10.1002/adfm.201402953

Sagane F, Shimokawa R, Sano H, Sakaebe H, Iriyama Y (2013a) In-situ scanning electron microscopy observations of Li plating and stripping reactions at the lithium phosphorus oxynitride glass electrolyte/Cu interface. J Power Sour 225:245–250. doi:10.1016/j.jpowsour.2012.10.026

Sagane F, Ikeda K-I, Okita K, Sano H, Sakaebe H, Iriyama Y (2013b) Effects of current densities on the lithium plating morphology at a lithium phosphorus oxynitride glass electrolyte/copper thin film interface. J Power Sour 233:34–42. doi:10.1016/j.jpowsour.2013.01.051

Sahu G, Lin Z, Li J, Liu Z, Dudney N, Liang C (2014) Air-stable, high-conduction solid electrolytes of arsenic-substituted Li$_4$SnS$_4$. Energy Environ Sci 7(3):1053–1058. doi:10.1039/c3ee43357a

Saint J, Morcrette M, Larcher D, Tarascon JM (2005) Exploring the Li–Ga room temperature phase diagram and the electrochemical performances of the LixGay alloys vs. Li. Solid State Ionics 176 (1–2):189–197. doi:10.1016/j.ssi.2004.05.021

Saito T, Uosaki K (2003) Surface film formation and lithium underpotential deposition on Au(111) surfaces in propylene carbonate. J Electrochem Soc 150(4):A532–A537. doi:10.1149/1.1557966

Saito K, Nemoto Y, Tobishima S, Yamaki J (1997) Improvement in lithium cycling efficiency by using additives in lithium metal. J Power Sour 68:476–479

Sakaebe H, Matsumoto H (2003) N-methyl-N-propylpiperidinium Bis(trifluoromethanesulfonyl) imide (PP13–TFSI)—Novel electrolyte base for Li battery. Electrochem Commun 5(7):594–598. doi:10.1016/s1388-2481(03)00137-1

Sakuda A, Hayashi A, Tatsumisago M (2010) Interfacial observation between LiCoO$_2$ electrode and Li$_2$S – P$_2$S$_5$ solid electrolytes of all-solid-state lithium secondary batteries using transmission electron microscopy. Chem Mater 22(3):949–956. doi:10.1021/cm901819c

Salomon M (1989) Electrolyte solvation in aprotic solvents. J Power Sour 26:9–21

Sangster J, Bale CW (1998) The Li–Sn (Lithium–Tin) system. J Phase Equilib 19(1):70–75

Sangster J, Pelton AD (1991a) The Li–W (Lithium–Tungsten) system. J Phase Equilib 12(2):203

Sangster J, Pelton AD (1991b) The Ga–Li (Gallium–Lithium) system. J Phase Equilib 12(1):33–36

Sangster J, Pelton AD (1991c) The In–Li (Indium–Lithium) system. J Phase Equilib 12(1):37–41

Sangster J, Pelton AD (1991d) The Li–Pt (Lithium–Platinum) system. J Phase Equilib 12(6):678–681

Sangster J, Pelton AD (1992) The Li–Pd (Lithium–Palladium) system. J Phase Equilib 13(1):63–66

Sano H, Sakaebe H, Matsumoto H (2011a) Observation of electrodeposited lithium by optical microscope in room temperature ionic liquid-based electrolyte. J Power Sour 196(16):6663–6669. doi:10.1016/j.jpowsour.2010.12.023

Sano H, Sakaebe H, Matsumoto H (2011b) Effect of organic additives on electrochemical properties of li anode in room temperature ionic liquid. J Electrochem Soc 158(3):A316–A321. doi:10.1149/1.3532054

Sano H, Sakaebe H, Matsumoto H (2012) In-situ optical microscope morphology observation of lithium electrodeposited in room temperature ionic liquids containing aliphatic quaternary ammonium cation. Electrochemistry 80 (10):777–779. doi:10.5796/electrochemistry.80.777

Sano H, Sakaebe H, Matsumoto H (2013) In situ morphology observations of electrodeposited lithium in room-temperature ionic liquids by optical microscopy. Chem Lett 42(1):77–79. doi:10.1246/cl.2013.77

Sano H, Sakaebe H, Senoh H, Matsumoto H (2014) Effect of current density on morphology of lithium electrodeposited in ionic liquid-based electrolytes. J Electrochem Soc 161(9):A1236–A1240. doi:10.1149/2.0331409jes

Sazhin SV, Gorodyskii AV, Khimchenko MY, Kuksenko SP, Danilin VV (1993) New parameters for lithium cyclability in organic electrolytes for secondary batteries. J Electroanal Chem 344:61–72

Sazhin SV, Gorodyskii AV, Khimchenko MY (1994) Lithium rechargeability on different substrates. J Power Sour 47:57–62

Sazhin SV, Khimchenko MY, Tritenichenko YN, Roh W, Kang HY (1997) Lithium state diagram as a description of lithium deposit morphology. J Power Sour 66:141–145

Schauser NS, Harry KJ, Parkinson DY, Watanabe H, Balsara NP (2014) Lithium dendrite growth in glassy and rubbery nanostructured block copolymer electrolytes. J Electrochem Soc 162 (3): A398–A405. doi:10.1149/2.0511503jes

Schechter A, Aurbach D, Cohen H (1999) X-ray photoelectron spectroscopy study of surface films formed on Li electrodes freshly prepared in alkyl carbonate solutions. Langmuir 15:3334–3342

Schedlbauer T, Krüger S, Schmitz R, Schmitz RW, Schreiner C, Gores HJ, Passerini S, Winter M (2013) Lithium Difluoro(oxalato)borate: a promising salt for lithium metal based secondary batteries? Electrochim Acta 92:102–107. doi:10.1016/j.electacta.2013.01.023

Scordilis-Kelley C, Affinito JD, Jones LD, Mikhaylik YV, Kovalev I, Wilkening WF, Campbell CTS, Martens JA (2015) Application of force in electrochemical cells. Appl Force Electrochem Cells US 9105938(B2):1–8

Seki S, Kobayashi Y, Miyashiro H, Ohno Y, Mita Y, Usami A, Terada N, Watanabe M (2005) Reversibility of lithium secondary batteries using a room-temperature ionic liquid mixture and lithium metal. Electrochem Solid-State Lett 8(11):A577–A578. doi:10.1149/1.2041330

Seki S, Kobayashi Y, Miyashiro H, Ohno Y, Usami A, Mita Y, Watanabe M, Terada N (2006a) Highly reversible lithium metal secondary battery using a room temperature ionic liquid/lithium salt mixture and a surface-coated cathode active material. Chem Commun (5):544–545. doi:10.1039/b514681j

Seki S, Kobayashi Y, Miyashiro H, Ohno Y, Usami A, Mita Y, Kihira N, Watanabe M, Terada N
    (2006b) Lithium secondary batteries using modified-imidazolium room-temperature ionic
    liquid. J Phys Chem B 110(21):10228–10230
Seki S, Ohno Y, Miyashiro H, Kobayashi Y, Usami A, Mita Y, Terada N, Hayamizu K, Tsuzuki S,
    Watanabe M (2008) Quaternary ammonium room-temperature ionic liquid/lithium salt binary
    electrolytes: electrochemical study. J Electrochem Soc 155(6):A421–A427. doi:10.1149/1.
    2899014
Selim R, Bro P (1974) Some observations on rechargeable lithium electrodes in a propylene
    carbonate electrolyte. J Electrochem Soc 121(11):1457–1459
Seong IW, Hong CH, Kim BK, Yoon WY (2008) The effects of current density and amount of
    discharge on dendrite formation in the lithium powder anode electrode. J Power Sour 178
    (2):769–773. doi:10.1016/j.jpowsour.2007.12.062
Shao Y, Ding F, Xiao J, Zhang J, Xu W, Park S, Zhang J-G, Wang Y, Liu J (2012b) Making
    Li-Air batteries rechargeable: material challenges. Adv Funct Mater:n/a–n/a. doi:10.1002/
    adfm.201200688
Shen DH, Subbarao S, Deligiannis F, Huang CK, Halpert G, Dominey LA, Koch VR, Goldman J
    (1991) Improved Lithium–Titanium disulfide cell cycling in ether-based electrolytes with
    synergistic additives: Part I. microcalorimetry, AC impedance spectroscopy and cell cycling
    studies In: Abraham KM, Salomon M (eds) Proceedings of the symposium on primary and
    secondary lithium batteries. The Electrochemical Society Inc, pp 280–292
Shin J-H, Henderson WA, Passerini S (2003) Ionic liquids to the rescue? Overcoming the ionic
    conductivity limitations of polymer electrolytes. Electrochem Commun 5 (12):1016–1020.
    doi:10.1016/j.elecom.2003.09.017
Shin J-H, Henderson WA, Passerini S (2005a) PEO-based polymer electrolytes with ionic liquids
    and their use in lithium metal-polymer electrolyte batteries. J Electrochem Soc 152(5):A978–
    A983. doi:10.1149/1.1890701
Shin J-H, Henderson WA, Passerini S (2005b) An elegant fix for polymer electrolytes.
    Electrochem Solid-State Lett 8(2):A125–A127. doi:10.1149/1.1850387
Shin J-H, Henderson WA, Scaccia S, Prosini PP, Passerini S (2006) Solid-state Li/LiFePO₄
    polymer electrolyte batteries incorporating an ionic liquid cycled at 40 °C. J Power Sour 156
    (2):560–566. doi:10.1016/j.jpowsour.2005.06.026
Shiraishi S, Kanamura K (1998) The observation of electrochemical dissolution of lithium metal
    using electrochemical quartz crystal microbalance and in-situ tapping mode atomic force
    microscopy. Langmuir 14:7082–7086
Shiraishi S, Kanamura K, Takehara Z-I (1995) Effect of surface modification using various acids
    on electrodeposition of lithium. J Appl Electrochem 25:584–591
Shiraishi S, Kanamura K, Takehara Z-I (1997) Study of the surface composition of highly smooth
    lithium deposited in various carbonate electrolytes containing HF. Langmuir 13:3542–3549
Shiraishi S, Kanamura K, Takehara Z-I (1999a) Influence of initial surface condition of lithium
    metal anodes on surface modification with HF. J Appl Electrochem 29:869–881
Shiraishi S, Kanamura K, Zi Takehara (1999b) Surface condition changes in lithium metal
    deposited in nonaqueous electrolyte containing hf by dissolution-deposition cycles.
    J Electrochem Soc 146(5):1633–1639
Shu J, Shui M, Huang F, Xu D, Ren Y, Hou L, Cui J, Xu J (2011) Comparative study on surface
    behaviors of copper current collector in electrolyte for lithium-ion batteries. Electrochim Acta
    56(8):3006–3014. doi:10.1016/j.electacta.2011.01.004
Singh M, Odusanya O, Wilmes GM, Eitouni HB, Gomez ED, Patel AJ, Chen VL, Park MJ,
    Fragouli P, Iatrou H, Hadjichristis N, Cookson D, Balsara NP (2007) Effect of molecular
    weight on the mechanical and electrical properties of block copolymer electrolytes.
    Macromolecules 40:4578–4585
Song JH, Yeon JT, Jang JY, Han JG, Lee SM, Choi NS (2013a) Effect of fluoroethylene carbonate
    on electrochemical performances of lithium electrodes and lithium-sulfur batteries.
    J Electrochem Soc 160(6):A873–A881. doi:10.1149/2.101306jes

Song J, Lee H, Choo MJ, Park JK, Kim HT (2015) Ionomer-liquid electrolyte hybrid ionic conductor for high cycling stability of lithium metal electrodes. Sci Rep 5:14458. doi:10.1038/srep14458

Stark JK, Ding Y, Kohl PA (2011) Dendrite-free electrodeposition and reoxidation of lithium-sodium alloy for metal-anode battery. J Electrochem Soc 158(10):A1100–A1105. doi:10.1149/1.3622348

Stark JK, Ding Y, Kohl PA (2013) Nucleation of electrodeposited lithium metal: dendritic growth and the effect of co-deposited sodium. J Electrochem Soc 160(9):D337–D342. doi:10.1149/2.028309jes

Stone GM, Mullin SA, Teran AA, Hallinan DT, Minor AM, Hexemer A, Balsara NP (2012) Resolution of the modulus versus adhesion dilemma in solid polymer electrolytes for rechargeable lithium metal batteries. J Electrochem Soc 159(3):A222–A227. doi:10.1149/2.030203jes

Sudha V, Sangaranarayanan MV (2002) Underpotential deposition of metals: structural and thermodynamic considerations. J Phys Chem B 106:2699–2707

Sudo R, Nakata Y, Ishiguro K, Matsui M, Hirano A, Takeda Y, Yamamoto O, Imanishi N (2014) Interface behavior between garnet-type lithium-conducting solid electrolyte and lithium metal. Solid State Ionics 262:151–154. doi:10.1016/j.ssi.2013.09.024

Suo L, Hu YS, Li H, Armand M, Chen L (2013) A new class of solvent-in-salt electrolyte for high-energy rechargeable metallic lithium batteries. Nat Commun 4:1481. doi:10.1038/ncomms2513

Surampudi S, Shen DH, Huang C-K, Narayanan SR, Attia A, Halpert G, Peled E (1993) Effect of cycling on the lithium/electrolyte interface in organic electrolytes. J Power Sour 43–44:21–26

Suresh P, Shukla AK, Shivashankar SA, Munichandraiah N (2002) Electrochemical behaviour of aluminium in non-aqueous electrolytes over a wide potential range. J Power Sour 110:11–18

Suzuki Y, Kami K, Watanabe K, Watanabe A, Saito N, Ohnishi T, Takada K, Sudo R, Imanishi N (2015) Transparent cubic garnet-type solid electrolyte of Al₂O₃-doped Li₇La₃Zr₂O₁₂. Solid State Ionics 278:172–176. doi:10.1016/j.ssi.2015.06.009

Swiderska-Mocek A, Naparstek D (2014) Compatibility of polymer electrolyte based on N-methyl-N-propylpiperidinium Bis(trifluoromethanesulphonyl)imide Ionic Liquid with LiMn₂O₄ cathode in Li–Ion batteries. Solid State Ionics 267:32–37. doi:10.1016/j.ssi.2014.09.007

Tachikawa H (1993) Characterization of lithium electrode surface in lithium secondary batteries by in situ Raman spectrscopic methods. US Government report AD-A263 728

Takada K (2013) Progress and prospective of solid-state lithium batteries. Acta Mater 61(3):759–770. doi:10.1016/j.actamat.2012.10.034

Takehara Z-I (1997) Future prospects of the lithium metal anode. J Power Sour 68:82–86

Takehara Z-I, Ogumi Z, Uchimoto Y, Yasuda K, Yoshida H (1993) Modification of lithium/electrolyte interface by plasma polymerization of 1,1-Difluoroethene. J Power Sour 43–44:377–383

Tatsumisago M, Mizuno F, Hayashi A (2006) All-solid-state lithium secondary batteries using sulfide-based glass–ceramic electrolytes. J Power Sour 159(1):193–199. doi:10.1016/j.jpowsour.2006.04.037

Tavassol H, Chan MKY, Catarello MG, Greeley J, Cahill DG, Gewirth AA (2013) Surface coverage and sei induced electrochemical surface stress changes during Li deposition in a model system for Li-Ion battery anodes. J Electrochem Soc 160(6):A888–A896. doi:10.1149/2.068306jes

Teran AA, Balsara NP (2011) Effect of lithium polysulfides on the morphology of block copolymer electrolytes. Macromolecules 44(23):9267–9275. doi:10.1021/ma202091z

Teran AA, Mullin SA, Hallinan DT, Balsara NP (2012) Discontinuous changes in ionic conductivity of a block copolymer electrolyte through an order–disorder transition. ACS Macro Lett 1:305–309. doi:10.1021/mz200183t

Thangadurai V, Weppner W (2005) Li6ALa2Ta2O12 (A = Sr, Ba): Novel Garnet-like oxides for fast lithium ion conduction. Adv Funct Mater 15(1):107–112

Thangadurai V, Weppner W (2006) Recent progress in solid oxide and lithium ion conducting electrolytes research. Ionics 12(1):81–92. doi:10.1007/s11581-006-0013-7

Thangadurai V, Kaack H, Weppner WJF (2003) Novel fast lithium ion conduction in garnet-type $Li_5La_3M_2O_{12}$ (M = Nb, Ta). J Am Ceram Soc 86(3):437–440

Tobishima S-I, Yamaki J-I, Yamaji A, Okada T (1984) Dialkoxyethane-propylene carbonate mixed electrolytes for lithium secondary batteries. J Power Sour 13:261–271

Tobishima S, Arakawa M, Hirai T, Yamaki J (1989) ethylene carbonate-based electrolytes for rechargeable lithium batteries. J Power Sour 26:449–454

Tobishima S, Arakawa M, Yamaki J (1990) Ethylene carbonate/linear-structured solvent mixed electrolyte systems for high-rate secondary lithium batteries. Electrochim Acta 35(2):383–388

Tobishima S-I, Hayashi K, Saito K-I, Yamaki J-I (1995) Ethylene carbonate-based ternary mixed solvent electrolytes for rechargeable lithium batteries. Electrochim Acta 40(5):537–544

Togasaki N, Momma T, Osaka T (2014) Enhancement effect of trace H_2O on the charge–discharge cycling performance of a Li metal anode. J Power Sour 261:23–27. doi:10.1016/j.jpowsour.2014.03.040

Uchiyama M, Slane S, Plichta E, Salomon M (1987) Solvent effects on rechargeable lithium cells. J Power Sour 20:279–286

Ueno M, Imanishi N, Hanai K, Kobayashi T, Hirano A, Yamamoto O, Takeda Y (2011) Electrochemical properties of cross-linked polymer electrolyte by electron beam irradiation and application to lithium ion batteries. J Power Sour 196(10):4756–4761. doi:10.1016/j.jpowsour.2011.01.054

van der Marel C, Vinkle GJB, Hennephof J, van der Lugt W (1982) The phase diagram of the system Lithium–Cadmium. J Phys Chem Solids 43(10):1013–1014

Vega JA, Zhou J, Kohl PA (2009) Electrochemical comparison and deposition of lithium and potassium from phosphonium- and Ammonium-TFSI ionic liquids. J Electrochem Soc 156(4):A253–A259. doi:10.1149/1.3070657

Venkatasetty HV (1975) Transport behavior and Raman spectra of electrolytes in methyl formate and propylene carbonate. J Electrochem Soc 122(2):245–249

Verma P, Maire P, Novák P (2010) A review of the features and analyses of the solid electrolyte interphase in Li–Ion batteries. Electrochim Acta 55(22):6332–6341. doi:10.1016/j.electacta.2010.05.072

Visco SJ, Nimon E, De Jonghe LC, Katz B, Chu MY (2004a) LITHIUM FUEL CELLS. Paper presented at the proceedings of the 12th international meeting on lithium batteries, June 27–July 2, 2004, Nara, Japan

Visco SJ, Nimon E, De Jonghe C (2009) Secondary batteries-metal-air systems: Lithium-Air. In: Garche J, Dyer C, Moseley P, Ogumi Z, Rand D, Scrosati B (eds) Encyclopedia of electrochemical power sources, vol 5. Elsevier, Amsterdam, pp 376–383

Wagner D, Gerischer H (1989) the deposition and reoxidation of lithium on gold in the underpotential range from acetonitrile solutions with small amounts of water. Electrochim Acta 34(9):1351–1356

Wainwright D, Shimizu R (1991) Forces generated by anode growth in cylindrical Li/MoS_2 Cells. J Power Sour 34:31–38

Wanakule NS, Panday A, Mullin SA, Gann E, Hexemer A, Balsara NP (2009) Ionic conductivity of block copolymer electrolytes in the vicinity of order–disorder and order–order transitions. Macromolecules 42(15):5642–5651. doi:10.1021/ma900401a

Wang J, King P, Huggins RA (1986) Investigations of binary Lithium–Zinc, Lithium–Cadmium and Lithium–Lead alloys as negative electrodes in organic solvent-based electrolyte. Solid State Ionics 20:185–189

Wang X, Yasukawa E, Mori S (1999) Electrochemical behavior of lithium imide/cyclic ether electrolytes for 4 V lithium metal rechargeable batteries. J Electrochem Soc 146:3992–3998

Wang X, Yasukawa E, Kasuya S (2000) Lithium imide electrolytes with two-oxygen-atom-containing cycloalkane solvents for 4 V lithium metal rechargeable batteries. J Electrochem Soc 147(7):2421–2426

Wang Y, Nakamura S, Ue M, Balbuena PB (2001) Theoretical studies to understand surface chemistry on carbon anodes for Lithium-Ion batteries: reduction mechanisms of ethylene carbonate. J Am Chem Soc 123:11708–11718

Wang C, Wang D, Wang Q, Chen H (2010) Fabrication and lithium storage performance of three-dimensional porous NiO as anode for Lithium-Ion battery. J Power Sour 195(21):7432–7437. doi:10.1016/j.jpowsour.2010.04.090

Wang H, Imanishi N, Hirano A, Takeda Y, Yamamoto O (2012) Electrochemical properties of the polyethylene oxide–Li(CF₃SO₂)2 N and ionic liquid composite electrolyte. J Power Sour 219:22–28. doi:10.1016/j.jpowsour.2012.07.020

Wang H, Im D, Lee DJ, Matsui M, Takeda Y, Yamamoto O, Imanishi N (2013a) A composite polymer electrolyte protect layer between lithium and water stable ceramics for aqueous Lithium-Air batteries. J Electrochem Soc 160(4):A728–A733. doi:10.1149/2.020306jes

Wang X, Hou Y, Zhu Y, Wu Y, Holze R (2013c) An aqueous rechargeable lithium battery using coated Li metal as anode. Sci Rep 3:1401. doi:10.1038/srep01401

Wang H, Zong Y, Zhao W, Sun L, Xin L, Liu Y (2015b) Synthesis of high aspect ratio CuO submicron rods through oriented attachment and their application in Lithium-Ion batteries. RSC Adv 5(62):49968–49972. doi:10.1039/c5ra07592k

Wang Y, Richards WD, Ong SP, Miara LJ, Kim JC, Mo Y, Ceder G (2015c) Design principles for solid-state lithium superionic conductors. Nat Mater 14(10):1026–1031. doi:10.1038/nmat4369

Wen CJ, Huggins RA (1980) Chemical diffusion in intermediate phases in the lithium–tin system. J Solid St Chem 35:376–384

Wen CJ, Huggins RA (1981) Chemical diffusion in intermediate phases in the Lithium–Silicon system. J Solid St Chem 37:271–278

Wen CJ, Boukamp BA, Huggins RA, Weppner W (1979) Thermodynamic and mass transport properties of "LiAl". J Electrochem Soc 126(12):2258–2266

Wetjen M, Kim G-T, Joost M, Winter M, Passerini S (2013) Temperature dependence of electrochemical properties of cross-linked Poly(ethylene oxide)–Lithium Bis(trifluoromethane-sulfonyl)imide–N-butyl-N-methylpyrrolidinium Bis(trifluoromethanesulfonyl)imide solid polymer electrolytes for lithium batteries. Electrochim Acta 87:779–787. doi:10.1016/j.electacta.2012.09.034

Wibowo R, Jones SEW, Compton RG (2009) Kinetic and thermodynamic parameters of the Li/Li + couple in the room temperature ionic liquid N-butyl-N-methylpyrrolidinium Bis (trifluoromethylsulfonyl) Imide in the temperature range 298–318 K: a theoretical and experimental study using Pt and Ni electrodes. J Phys Chem B 113:12293–12298

Wibowo R, Jones SEW, Compton RG (2010) Investigating the electrode kinetics of the Li/Li + couple in a wide range of room temperature ionic liquids at 298 K. J Chem Eng Data 55:1374–1376

Wilkinson DP, Wainwright D (1993) Control of lithium metal anode cycleability by electrolyte temperature. J Electroanal Chem 355:193–203

Wilkinson DP, Blom H, Brandt K, Wainwright D (1991) Effects of physical constraints on Li cyclability. J Power Sour 36:517–527

Wu M, Wen Z, Liu Y, Wang X, Huang L (2011) Electrochemical behaviors of a Li3 N modified Li metal electrode in secondary lithium batteries. J Power Sour 196(19):8091–8097. doi:10.1016/j.jpowsour.2011.05.035

Wu M, Wen Z, Jin J, Cui Y (2013) Effects of combinatorial AlCl₃ and pyrrole on the SEI formation and electrochemical performance of Li electrode. Electrochim Acta 103:199–205. doi:10.1016/j.electacta.2013.03.181

Wu F, Qian J, Chen R, Lu J, Li L, Wu H, Chen J, Zhao T, Ye Y, Amine K (2014) An effective approach to protect lithium anode and improve cycle performance for Li-S batteries. ACS Appl Mater Interfaces 6(17):15542–15549. doi:10.1021/am504345s

Xianming W, Yasukawa E, Kasuya S (2001) electrochemical properties of tetrahydropyran-based ternary electrolytes for 4 V lithium metal rechargeable batteries. Electrochim Acta 46:813–819

Xiong S, Kai X, Hong X, Diao Y (2011) Effect of LiBOB as additive on electrochemical properties of Lithium–Sulfur batteries. Ionics 18(3):249–254. doi:10.1007/s11581-011-0628-1

Xu X, Han M, Ma J, Zhang C, Li G (2015) Preparation of a nanoporous CuO/Cu composite using a dealloy method for high performance Lithium-Ion batteries. RSC Adv 5(88):71760–71764. doi:10.1039/c5ra14123k

Xue Z, He D, Xie X (2015) Poly(ethylene oxide)-based electrolytes for Lithium-Ion batteries. J Mater Chem A 3(38):19218–19253. doi:10.1039/c5ta03471j

Yamada Y, Takazawa Y, Miyazaki K, Abe T (2010) Electrochemical lithium intercalation into graphite in dimethyl sulfoxide-based Electrolytes: effect of solvation structure of lithium ion. J Phys Chem C 114:11680–11685

Yamada Y, Yaegashi M, Abe T, Yamada A (2013) A superconcentrated ether electrolyte for fast-charging Li-Ion batteries. Chem Commun 49(95):11194–11196. doi:10.1039/c3cc46665e

Yamada Y, Usui K, Chiang CH, Kikuchi K, Furukawa K, Yamada A (2014a) General observation of lithium intercalation into graphite in ethylene-carbonate-free superconcentrated electrolytes. ACS Appl Mater Interfaces 6(14):10892–10899. doi:10.1021/am5001163

Yamada Y, Furukawa K, Sodeyama K, Kikuchi K, Yaegashi M, Tateyama Y, Yamada A (2014b) Unusual stability of acetonitrile-based superconcentrated electrolytes for fast-charging Lithium-Ion batteries. J Am Chem Soc 136(13):5039–5046

Yamamoto K, Iriyama Y, Asaka T, Hirayama T, Fujita H, Nonaka K, Miyahara K, Sugita Y, Ogumi Z (2012) Direct observation of lithium-ion movement around an in-situ-formed-negative-electrode/solid-state-electrolyte interface during initial charge–discharge reaction. Electrochem Commun 20:113–116. doi:10.1016/j.elecom.2012.04.013

Yang H, Fey EO, Trimm BD, Dimitrov N, Whittingham MS (2014) Effects of pulse plating on lithium electrodeposition, morphology and cycling efficiency. J Power Sour 272:900–908. doi:10.1016/j.jpowsour.2014.09.026

Yoo E, Zhou H (2011) Li−Air rechargeable battery based on metal-free graphene nanosheet catalysts. ACS Nano 5(4):3020–3026

Yoon S, Lee J, Kim S-O, Sohn H-J (2008) Enhanced cyclability and surface characteristics of lithium batteries by Li–Mg Co-deposition and addition of HF acid in electrolyte. Electrochim Acta 53(5):2501–2506. doi:10.1016/j.electacta.2007.10.019

Yoshida K, Nakamura M, Kazue Y, Tachikawa N, Tsuzuki S, Seki S, Dokko K, Watanabe M (2011) Oxidative-stability enhancement and charge transport mechanism in glyme-lithium salt equimolar complexes. J Am Chem Soc 133(33):13121–13129. doi:10.1021/ja203983r

Yoshimatsu I, Hirai T, Yamaki J-I (1988) Lithium electrode morphology during cycling in lithium cells. J Electrochem Soc 135(10):2422–2427

Young W-S, Epps TH (2012) Ionic conductivities of block copolymer electrolytes with various conducting pathways: sample preparation and processing considerations. Macromolecules 45 (11):4689–4697. doi:10.1021/ma300362f

Young W-S, Brigandi PJ, Epps TH III (2008) Crystallization-induced lamellar-to-lamellar thermal transition in salt-containing block copolymer electrolytes. Macromolecules 41:6276–6279

Young W-S, Albert JNL, Schantz AB, Epps TH (2011) Mixed-salt effects on the ionic conductivity of lithium-doped PEO-containing block copolymers. Macromolecules 44 (20):8116–8123. doi:10.1021/ma2013157

Young W-S, Kuan W-F, Epps TH (2014) Block copolymer electrolytes for rechargeable lithium batteries. J Polym Sci B 52(1):1–16. doi:10.1002/polb.23404

Yuan WX, Wang WJ, Song YT, Chen XL (2003) Thermodynamic descriptions of the Ga–Li system. Scripta Mater 48(8):1053–1059. doi:10.1016/s1359-6462(02)00634-6

Yun YS, Kim JH, Lee S-Y, Shim E-G, Kim D-W (2011) Cycling performance and thermal stability of lithium polymer cells assembled with ionic liquid-containing gel polymer electrolytes. J Power Sour 196(16):6750–6755. doi:10.1016/j.jpowsour.2010.10.088

Zaban A, Aurbach D (1995) impedance spectroscopy of lithium and nickel electrodes in propylene carbonate solutions of different lithium salts. Comp Study J Power Sour 54:289–295

Zaban A, Zinigrad E, Aurbach D (1996) Impedance spectroscopy of Li electrodes. 4. A general simple model of the Li−solution interphase in polar aprotic systems. J Phys Chem 100:3089–3101

Zeng Z, Liang WI, Chu YH, Zheng H (2014) In situ TEM study of the Li–Au reaction in an electrochemical liquid cell. Faraday Discuss 176:95–107. doi:10.1039/c4fd00145a

Zhang SS (2013a) New insight into liquid electrolyte of rechargeable lithium/sulfur battery. Electrochim Acta 97:226–230. doi:10.1016/j.electacta.2013.02.122

Zhang X-W, Wang C, Appleby AJ, Little FE (2002) Characteristics of Lithium-Ion-conducting composite polymer-glass secondary cell electrolytes. J Power Sour 112:209–215

Zhang J, Xie S, Wei X, Xiang YJ, Chen CH (2004a) Lithium insertion in naturally surface-oxidized copper. J Power Sour 137(1):88–92. doi:10.1016/j.jpowsour.2004.05.041

Zhang X-W, Li Y, Khan SA, Fedkiw PS (2004b) Inhibition of lithium dendrites by fumed silica-based composite electrolytes. J Electrochem Soc 151(8):A1257-A1263. doi:10.1149/1.1767158

Zhang T, Imanishi N, Hasegawa S, Hirano A, Xie J, Takeda Y, Yamamoto O, Sammes N (2008) Li/polymer electrolyte/water stable lithium-conducting glass ceramics composite for lithium–air secondary batteries with an aqueous electrolyte. J Electrochem Soc 155(12):A965–A969. doi:10.1149/1.2990717

Zhang T, Imanishi N, Hasegawa S, Hirano A, Xie J, Takeda Y, Yamamoto O, Sammes N (2009) Water-stable lithium anode with the three-layer construction for aqueous lithium–air secondary batteries. Electrochem Solid-State Lett 12(7):A132–A135. doi:10.1149/1.3125285

Zhang T, Imanishi N, Shimonishi Y, Hirano A, Takeda Y, Yamamoto O, Sammes N (2010b) A novel high energy density rechargeable lithium/air battery. Chem Commun 46(10):1661–1663. doi:10.1039/b920012f

Zhang T, Imanishi N, Shimonishi Y, Hirano A, Xie J, Takeda Y, Yamamoto O, Sammes N (2010c) Stability of a water-stable lithium metal anode for a lithium–air battery with acetic acid–water solutions. J Electrochem Soc 157(2):A214-A218. doi:10.1149/1.3271103

Zhang T, Imanishi N, Hirano A, Takeda Y, Yamamoto O (2011) Stability of Li/polymer electrolyte-ionic liquid composite/lithium conducting glass ceramics in an aqueous electrolyte. Electrochem Solid-State Lett 14 (4):A45–A48. doi:10.1149/1.3545964

Zhang YJ, Liu XY, Bai WQ, Tang H, Shi SJ, Wang XL, Gu CD, Tu JP (2014c) Magnetron sputtering amorphous carbon coatings on metallic lithium: towards promising anodes for lithium secondary batteries. J Power Sour 266:43–50. doi:10.1016/j.jpowsour.2014.04.147

Zheng J, Qin J, Zhao Y, Abe T, Ogumi Z (2005) Temperature dependence of the electrochemical behavior of licoo2 in quaternary ammonium-based ionic liquid electrolyte. Solid State Ionics 176 (29–30):2219–2226. doi:10.1016/j.ssi.2005.06.020

Zheng G, Lee SW, Liang Z, Lee HW, Yan K, Yao H, Wang H, Li W, Chu S, Cui Y (2014) Interconnected hollow carbon nanospheres for stable lithium metal anodes. Nat Nanotechnol 9 (8):618–623. doi:10.1038/nnano.2014.152

Zheng J, Yan P, Mei D, Engelhard MH, Cartmell SS, Polzin BJ, Wang C, Zhang J-G, Wu Xu (2015) Highly stable operation of li metal batteries enabled by the formation of transient high concentration electrolyte layer. Submitted to advanced energy materials

Zhou YN, Wang XJ, Lee HS, Nam KW, Yang XQ, Haas O (2010) Electrochemical investigation of Li–Al anodes in Oligo(ethylene glycol) Dimethyl Ether/LiPF6. J Appl Electrochem 41 (3):271–275. doi:10.1007/s10800-010-0233-4

Zhu C, Cheng H, Yang Y (2008) Electrochemical characterization of two types of PEO-based polymer electrolytes with room-temperature ionic liquids. J Electrochem Soc 155(8):A569–A575. doi:10.1149/1.2931523

Zhuang G, Wang K, Chottiner G, Barbour R, Luo Y, Bae IT, Tyrk D, Scherson DA (1995) novel in situ and ex situ techniques for the study of lithium/electrolyte interfaces. J Power Sour 54:20–27

Zinigrad E, Levi E, Teller H, Salitra G, Aurbach D, Dan P (2004) Investigation of lithium electrodeposits formed in practical rechargeable Li-LixMnO$_2$ batteries based on LiAsF$_6$/ 1,3-Dioxolane solutions. J Electrochem Soc 151(1):A111–A118. doi:10.1149/1.1630591

Zlatilova P, Balkanov I, Geronov Y (1988) Thin foil Lithium–Aluminum electrode. The effect of thermal treatment on its electrochemical behavior in nonaqueous media. J Power Sour 24:71–79

Chapter 4
Application of Lithium Metal Anodes

Li metal is an ideal anode to replace Li intercalation or conversion compounds used in Li-ion batteries (Whittingham 2012). It is also widely used in Li–S (Bruce et al. 2012; Ji 2009) and Li-air batteries (Girishkumar 2010; Lee 2011), as well as other rechargeable Li metal batteries using Li intercalation or conversion compounds as cathode. Although the use of Li metal anodes in these batteries has been limited by Li dendrite growth and the low CE of Li cycling, the stability of Li metal anodes is much different when used in different types of Li metal batteries. For example, a Li metal anode is much more stable in sealed batteries (such as batteries using a Li intercalation compound (such as $LiCoO_2$ or $LiFePO_4$) or Li conversion compounds [such as S] as cathode) than in batteries that are open to the ambient air (such as Li-air batteries). On the other hand, a Li metal anode is more stable in an electrolyte suitable for Li intercalation compounds (such as $LiCoO_2$ or $LiFePO_4$) than in an electrolyte that is suitable for a sulfur-based cathode. Therefore, it will be important to analyze the stability and protection of Li metal anodes in different types of Li metal batteries.

4.1 Lithium Metal Batteries with Lithium Intercalation Cathodes

In a battery with a Li metal anode and a Li intercalation compound (such as $LiCoO_2$ or $LiFePO_4$) as cathode, most electrolytes used in conventional Li-ion batteries can be used. In this case, the main challenge for use of Li metal anodes is Li dendrite growth and eventual growth of a passivation layer (or SEI layer) formed by the chemical and electrochemical reactions between Li and the electrolyte. These side reactions not only result in low CE of Li metal anodes, they also lead to loss of electrolyte. In fact, battery dry-out is one of the main reasons for the sudden failure of Li metal and Li-ion batteries.

Exxon developed Li∥TiS_2 batteries in the 1970s and produced cells as large as 45 Wh. Initial cells used DOL as the electrolyte solvent because it allowed the

© Springer International Publishing Switzerland 2017
J.-G. Zhang et al., *Lithium Metal Anodes and Rechargeable
Lithium Metal Batteries*, Springer Series in Materials Science 249,
DOI 10.1007/978-3-319-44054-5_4

Fig. 4.1 Discharge/charge curve of Li‖TiS$_2$ battery at 10 mA cm^2. Reproduced with permission—Copyright 1978, Elsevier (Whittingham 2004)

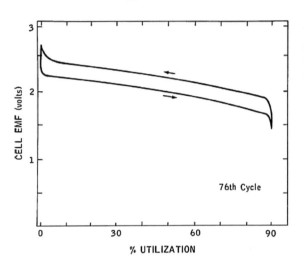

cycling of all of the Li in the Li anode for more than 100 cycles. Initially, lithium perchlorate (LiClO$_4$) was used as the electrolyte salt, but it was not stable. Later, the perchlorate was replaced by tetramethyl boron (CH$_3$)$_4$B$^-$ anion. Figure 4.1 shows a discharge/charge curve of a titanium disulfide (Li‖TiS$_2$) battery at 10 mA/cm^2 (Whittingham 2004). The electrolyte used was 2.5 M LiClO$_4$ in DOL, which was found to be an exceptional electrolyte for effective Li plating. The ready reversibility of Li in TiS$_2$ has permitted deep cycling for close to 1000 cycles with minimal capacity loss—less than 0.05 % per cycle—with excess Li anodes. However, the plating of Li metal is problematic, with dendrites being formed and led to battery shorts and even fires.

Hydro-Québec started to work on Li metal with polymer electrolyte in rechargeable Li batteries in 1980 (Zaghib 2012). Several new polymer and solid electrolytes with improved conductivity have resulted from a better understanding of the major parameters controlling the ion migration, such as favorable polymer structure, phase diagram between solvating polymer and Li salt and, the development of new counter-anions. This technology has been acquired and further developed for electric vehicle applications by Vehicule Électriques Pininfarina Bolloré (Wikipedia). Bolloré Bluecar used a 30 kWh lithium metal polymer (LMP) battery coupled to a supercapacitor, that provides an electric range of 250 km (160 mi) in urban use, and a maximum speed of 130 km h^{-1} (81 mph). The LMP batteries consist of a laminate of four ultra-thin materials: (1) metallic Li foil anode that acts as both a Li source and a current collector; (2) solid polymeric electrolyte created by dissolving a Li salt in a solvating copolymer (poly-oxyethylene); (3) cathode composed of vanadium oxide, carbon, and polymer to form a plastic composite; and (4) aluminum foil current collector. Figure 4.2 shows a Bluecar charged on the street of Paris.

Moli Energy developed Li‖molybdenum disulfide (MoS$_2$) batteries with excellent performance. However, the batteries were recalled in 1989 because of an

Fig. 4.2 A Bolloré Bluecar in Paris [Photo by: Mariordo (Mario Roberto Durán Ortiz)] from Wikipedia, the free encyclopedia (Wikipedia)

overheating defect. AT&T also experienced this when deploying Avestor batteries that used a Li anode, a vanadium oxide cathode, and a polymeric membrane (see Fig. 4.3). A rechargeable Li‖Li_xMnO_2 3 V battery was developed by Tadiran Ltd. At charge regimes around C/10, more than 350 cycles at 100 % DOD could be obtained (Tobishima et al. 1996). The safety hazard that occurred in a series of Li metal batteries from the middle 1970s to early 1990s has prevented

Fig. 4.3 The result of an Avestor battery fire in an AT&T system. Reproduced with permission—Copyright 2012, IEEE (Whittingham 2012)

commercialization of rechargeable Li metal batteries, and Li-ion batteries using graphite as a safe anode have dominated the market of Li-based rechargeable batteries since then.

4.2 Lithium Metal Anodes in Lithium–Sulfur Batteries

With the significant progresses made on the development of cathode materials and electrolytes in Li–S batteries in recent years, the stability of Li anode in Li–S batteries has become one of the more urgent challenges for reaching the long-term stability of Li–S batteries. In Li–S batteries, a passivation layer is easily formed on the metallic Li anode surface because of the presence of polysulfides and electrolyte additives. Although the passivation layer on a Li metal anode can significantly suppress Li dendrite growth and improve the safety of Li–S batteries, continuous corrosion of a Li metal anode eventually leads to battery failure due to the increased cell impedance and the depletion of electrolyte. Therefore, minimizing the corrosion of Li anodes is critical for the operation of Li–S batteries.

The Li–S battery has attracted tremendous attention from the energy storage community, mainly due to its high theoretical energy density (~ 2567 Wh kg^{-1}), which is more than twice that of conventional Li-ion batteries; this makes it one of the most promising candidates to meet the energy needs for powering future electric vehicles (Nazar et al. 2014; Bruce et al. 2012; Evers and Nazar 2012; Manthiram et al. 2012, 2014; Yin et al. 2013; Yang et al. 2013a; Xu et al. 2014a; Chen and Shaw 2014). In contrast to enormous research efforts on the S cathode side, the Li metal anode in Li–S batteries, which is directly involved in the shuttle mechanism and the capacity failure, has attracted much less attention (Manthiram et al. 2014; Lee et al. 2003). Recently, with the significant improvements in the development of high capacity S cathodes using stable C/S composites, the research spotlight is falling on the anode side which seems to be an equally major limiting factor for Li–S batteries as practical energy storage devices (Kulisch et al. 2014; Huang et al. 2014a; Yang et al. 2010; Zu and Manthiram 2014; Brückner et al. 2014; Elazari et al. 2012).

Metallic Li is a preferred electrode material for anodes of Li–S batteries. However, to use Li metal as the anode in nonaqueous electrolytes is a challenging task, because the Li anode forms dendritic and mossy metal deposits, leading to low CE, short cycle life, and safety concern (Liu et al. 2014a). In addition, unlike graphite and silicon (Si) anodes where lithiation leads to a volume increase of about 10 and 300 %, respectively, the relative volumetric change of a metallic Li anode is actually infinite (Zheng et al. 2014). To solve the aforementioned fundamental challenges facing metallic Li anodes, generally, two strategies have been employed to tackle the "holy grail" of rechargeable Li batteries: one strategy is to form in situ a stable SEI layer on the Li anode, which could level Li deposition and withstand mechanical deformation; another is to make an ex situ conformal coating on the Li anode or use a solid electrolyte layer to isolate Li from electrolytes (Younesi et al.

2013; Bryngelsson et al. 2007; Xiong et al. 2011, 2012, 2013, 2014a, b, c; Munichandraiah et al. 1998; Ratnakumar et al. 2001). This problem becomes more complicated in Li–S batteries because the presented polysulfides and additives in electrolytes inevitably react with Li metal, forming a passivation layer (Yamin and Peled 1983; Song et al. 2013a; Barchasz et al. 2013a, b; Park et al. 2013; Wu et al. 2014; Wang et al. 2014).

4.2.1 Performance and Characteristics of Lithium–Sulfur Batteries

In a Li–S battery, the passivation layer formed on the Li metal anode plays a significant role. In the typical ether-based electrolytes widely used in literature, such as 1 M LiTFSI-DOL/DME (1:1, v/v), polysulfides formed on the S cathode side during the discharge process could migrate to the Li anode side to form a passivation layer and subsequently shuttle back to the cathode upon charging; this is referred to as the shuttle mechanism, which leads to a low CE and a short cycle life of a Li–S battery. The polysulfides involved in this process lead to a more complicated SEI layer on the Li anode surface and make it more dynamic than that in conventional Li-ion batteries. The addition of $LiNO_3$ in the ether-based electrolytes could significantly improve the CE and the cycle life of Li–S batteries and, to some extent, mitigate the safety concern (Zhang 2012a, b, 2013b; Zhang and Read 2012; Liang et al. 2011b). The SEI layer formed on Li anodes in Li–S batteries has been characterized using XPS and SEM (Xiong et al. 2014a). This SEI film is assumed to consist of two sublayers (Fig. 4.4): the top layer is composed of oxidized products from polysulfides (mainly lithium sulfates) and the bottom layer is composed of the reduced products of polysulfides and $LiNO_3$ (lithium sulfides and LiN_xO_y) (Xiong et al. 2014a).

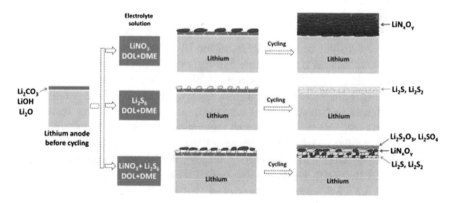

Fig. 4.4 Illustration of the surface film behavior on Li anode cycling in typical Li-S electrolytes. Reproduced with permission—Copyright 2014, Elsevier (Xiong et al. 2014a)

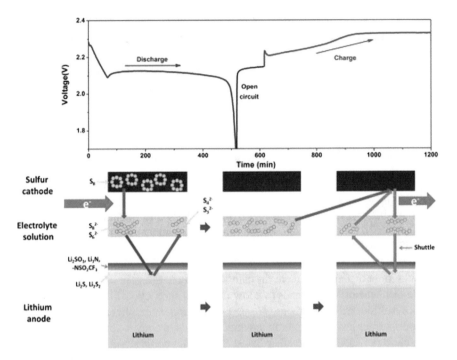

Fig. 4.5 Schematic representation of the surface film behavior on the Li anode in a Li-S battery cycled with 0.8 M LiTFSI/0.2 M Li$_2$S$_8$/DOL/DME (1:1, v/v). Reproduced with permission— Copyright 2013, Elsevier (Xiong et al. 2013)

The role of the passivation layer formed on the Li anode in Li–S batteries is still elusive. On the one hand, as shown in Fig. 4.5, the formation of the passivation layer during discharge could increase the impedance, leading to a high internal resistance of the cell. The passivation layer could also be broken in the following charge process, leading to a very dynamic feature in the SEI evolution (Xiong et al. 2013). On the other hand, the passivation layer could block further attacks from polysulfides and reduce the degradation of the Li anode. In a solvent-in-salt (i.e., concentrated) electrolyte, a more stable SEI layer could be formed on a metallic Li anode than that in a conventional electrolyte with the salt concentration around 1 M (Suo et al. 2013). In such a concentrated electrolyte, the Li anode exhibited the lowest roughness and damage level compared with other Li anodes in low salt concentration electrolytes, indicating the prevention of dendritic Li growth during cycling. A protective passivation layer, which was formed on the surface of the Li anode in a Li–S battery through simply soaking the Li strip in an electrolyte of 0.5 M LiTFSI in a DOL/DME solution with 1 wt% LiNO$_3$, could enable a good cycle life of the Li–S batteries (Yang et al. 2013b). It was also reported that a protective SEI layer formed in situ that was created by placing S powders in contact with the Li anode showed great benefit for the cell cycling performance (Demir-Cakan et al. 2013). XPS analysis indicated that the in situ formed SEI layer contained more long-chain Li_2S_x and less Li_2S compared to the SEI formed in conventional Li–S batteries, which should contribute to the improvement of cell performance. In addition, temperature also significantly affects the kinetics of a Li–S battery and also influences the formation of the SEI layer on the Li anode (Busche et al. 2014). A thicker and more stable SEI layer was formed on the Li anode surface at higher temperatures, which could prevent polysulfide diffusion and dendrite formation on the Li anode (Kim et al. 2013a).

The passivation layer formed on the Li anode in a polysulfide-rich electrolyte is not dense and rigid enough to totally block polysulfides and eliminate the shuttle effect. In particular, Li_2S_x in the passivation layer may decompose during the charge process, and as a result the passivation layer would be deconstructed, which could lead to a fast degradation of the Li anode. It is widely believed that polysulfides are able to penetrate the passivation layer and continue corroding the fresh Li source underlining the surface layer, leading to capacity loss (Fu et al. 2013). The presence of polysulfides in the electrolyte of a Li–S battery could also result in a higher SEI layer resistance and a higher activation energy (Xiong et al. 2014b). Therefore, the electrolyte volume contained in the Li–S battery is a more important factor than that in conventional Li-ion batteries. With the decrease of the relative amount of electrolyte (increasing S/electrolyte or S/E ratio in Fig. 4.6), the morphology of a cycled Li metal anode became closer to that of fresh Li in terms of both surface and cross-section views, indicating a better passivation layer caused by the higher concentration of polysulfides in the electrolyte (Zheng et al. 2013a). In the next

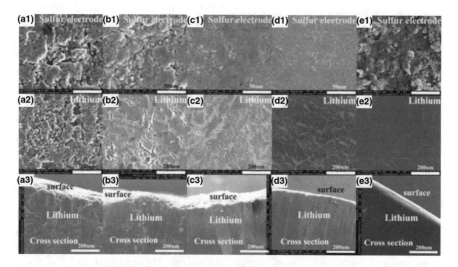

Fig. 4.6 SEM images of sulfur electrodes cycled in cells with different S/E ratios: *a1* 15 g L^{-1}, *b1* 20 g L^{-1}, *c1* 50 g L^{-1}, *d1* 75 g L^{-1}, and *e1* fresh sulfur electrode. Surface and cross-section SEM images for Li anodes cycled in cells with different S/E ratios: *a2, a3* 15 g L^{-1}, *b2, b3* 20 g L^{-1}, *c2, c3* 50 g L^{-1}, *d2, d3* 75 g L^{-1}; and *e2, e3* fresh Li anode. Reproduced with permission—Copyright 2013, The Electrochemical Society (Zheng et al. 2013a)

section, we will introduce some key strategies to handle the passivation layers formed on Li metal anodes in Li–S batteries.

Corrosion of the Li anode, which is caused by side reactions of the components dissolved in electrolytes and Li metal, is one of the main characteristics leading to the cell failure mechanism of rechargeable Li batteries, including Li–S batteries (Aurbach et al. 2000). To effectively protect the Li metal from corrosion/side reactions, a deep understanding of the degradation mechanism of the Li anode is necessary. The degradation of the Li anode in a Li–S battery is much more complicated than that in conventional Li metal batteries that using Li intercalation compounds as cathode because of the presence of polysulfides. Even in conventional electrolytes, the formation of the SEI and its degradation mechanism are still elusive due to the complicated interfacial chemistry between the Li metal and the electrolyte. The SEI chemistry and the evolution of Li electrode upon cycling are probably the most important questions that must be answered, but have not been satisfactorily addressed to date. Lv et al. (2015) tried to answer these questions by employing NCA as a model cathode of Li-ion battery cathode and a Li metal as the anode. Surprisingly, it was found that the quick growth of a porous SEI toward the bulk Li, rather than dendrites growing outward from the metal surface toward/through the separator, dramatically built up the cell impedance, which directly led to the onset of cell degradation and failure (Fig. 2.21).

A straightforward strategy to prevent corrosion of the Li anode in a Li–S battery is the use of a physical barrier to block polysulfides shuttling back and forth between the S cathode and Li anode. Solid-state electrolytes or polymer electrolytes have been intensively investigated to block polysulfides and were proven to largely eliminate the shuttle effect (Nagao et al. 2012b, 2013b; Agostini et al. 2013; Lin et al. 2013a; Kinoshita et al. 2014; Unemoto et al. 2014; Zhang et al. 2014; Jeddi et al. 2013; Jin et al. 2013b; Zhang 2013c; Zhang and Tran 2013; Jeddi et al. 2014; Vaughey et al. 2014; Bucur et al. 2013). However, the low ionic conductivity of solid-state and polymer electrolytes at room temperature limits their wide application in Li–S batteries. Therefore, an alternative and more practical approach, which uses the conventional ether-based liquid electrolytes along with an ion-selective membrane, has been developed, where the high ionic conductivity of the liquid electrolyte and the ionic selectivity of the separator are maintained. For instance, the traditional porous polyolefin separators were modified with Nafion polymer molecules containing SO_3^- groups that could reject the hopping of negative ions, such as polysulfide anions, due to Coulombic interactions (Fig. 4.7) (Huang et al. 2014b). Using this strategy, an ultralow capacity loss of 0.08 % per cycle was achieved within 500 cycles (Huang et al. 2014b). A functional separator with lithium perfluorinated sulfonyl dicyanomethide (Li-PFSD) was also investigated in Li–S batteries and showed even better rate performance compared to the battery with the lithiated Nafion ionomer electrolyte (Jin et al. 2013a).

Physical coating with inorganic or polymeric materials on either the Li metal or the cathode have proven to be a successful strategy to significantly block the migration of polysulfides from the S cathode to the Li anode, and accordingly mitigate the passivation of the Li anode (Yu et al. 2014; Ma et al. 2014; Oh et al.

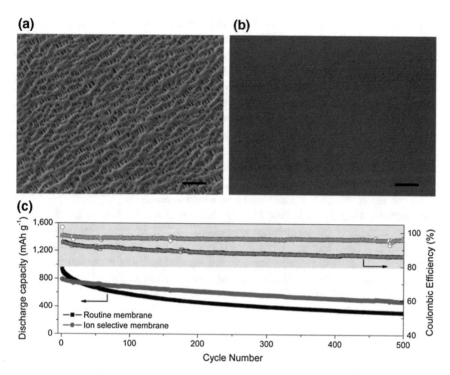

Fig. 4.7 The morphologies of the routine and ion-selective membranes: **a** the SEM image of routine PP/PE/PP membrane and **b** ion-selective Nafion–PP/PE/PP membrane. Scale bar: 1 μm. **c** Cyclic and rate performance of cells. Reproduced with permission—Copyright 2014, The Royal Society of Chemistry (Huang et al. 2014b)

2014; Tang et al. 2014; Liu et al. 2014b; Huang et al. 2013; Oh and Yoon 2014). Particularly, the insertion of an interlayer between the S cathode and the separator has been extensively explored (Jeong et al. 2013; Su and Manthiram 2012; Chung and Manthiram 2014a, b, c; Zu et al. 2013; Li et al. 2014a). Manthiram and coworkers investigated a variety of carbon types, including an eggshell membrane (Chung and Manthiram 2014a), a microporous carbon paper (Su and Manthiram 2012), a carbonized leaf (Chung and Manthiram 2014b), a carbon-coated separator (Chung and Manthiram 2014d), a carbon nanotube-coated separator (Chung and Manthiram 2014c), and a treated carbon paper-based interlayer (Zu et al. 2013) as a barrier to block the polysulfide shuttle. A graphene layer coated on the separator was also tested in order to mitigate the polysulfide migration from the cathode to the Li anode; this enabled stable cycling up to 300 cycles (Zhou et al. 2014). Wang et al. demonstrated that the reduced graphene layer combined with a conductive carbon black could improve the blocking effect and result in better cell performance (Wang et al. 2013b). A conductive coating on the separator of a thin layer of carbon materials could not only prevent the shuttle effect but also improve the utilization of the polysulfides accommodated in the separator (Yao et al. 2014).

It should be noted that the above strategies of using ion-selective membranes and carbon interlayers in the literature reports seemed more focused on improving the utilization of the S cathodes and intercepting the migration of polysulfides, but were not directly related to the protection of Li metal anodes. However, as discussed above, the existence of polysulfides during discharge/charge processes of the Li–S battery makes the electrolyte system more complicated than that in conventional Li-ion batteries. In fact, the attack of polysulfides on metallic Li in Li–S batteries always leads to quick degradation of a Li anode. Therefore, the introduction of such ion-selective membranes and/or carbon interlayers can not only improve the S utilization in cathode, but also protect the Li anode by blocking the migration of polysulfides.

A different approach to protect the Li anode from attack by polysulfides has been developed by Huang et al. (Figure 4.8) (Huang et al. 2014a). A hybrid anode structure with graphite electrically connected to the Li metal in parallel was employed in a Li–S battery. The lithiated graphite functioned as an artificial SEI layer on Li metal that supplies the Li^+ ion on demand while minimizing direct contact between soluble polysulfides and the Li metal surface. The Li–S cells incorporating the hybrid anodes delivered capacities >800 mAh g^{-1} for 400 cycles at a high rate of 1737 mA g^{-1}, with a CE >99 % and only 11 % capacity fade.

Fig. 4.8 Schematic of the hybrid anode design to manipulate the surface reactions on Li–S batteries. Reproduced with permission —Copyright 2014, Macmillan Publishers Ltd. (Huang et al. 2014a)

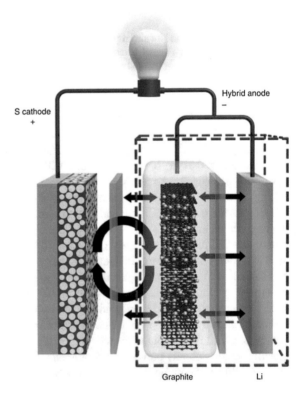

Besides insertion of a physical barrier between the S cathode and the Li anode, a conformal coating of inorganic or organic materials on the Li metal surface is also an effective strategy to prevent corrosion or side reactions of the Li metal from forming a passivation layer. One example of such coating layers was formed by a crosslinking reaction of a curable monomer in the presence of liquid electrolyte and a photo-initiator (Lee et al. 2003). With the protection layer, an enhanced cell performance was achieved in comparison with cells using a polymer electrolyte. In another work, conductive polymers, such as polypyrrole, have been used to form a protective coating layer on Li powder for Li–S batteries; it can inhibit the side reactions between the Li powder anode and dissolved polysulfides in the electrolyte, minimizing the overcharge and reducing the capacity fading (Oh and Yoon 2014). Similarly, it has been reported that Li powders with protective coatings, compared to Li metal foil, exhibit more advantages, including dendrite suppression, leading to a safer manufacturing process and safer cell operation (Heine et al. 2014). Kim et al. (2013b) reported that a coating layer of a Li–Al alloy could mitigate the polysulfide shuttle phenomenon in Li–S batteries. A protective film consisting of a Li–Al alloy and lithium pyrrolide film was formed on the Li electrode in an electrolyte containing pyrrole and $AlCl_3$ (Wu et al. 2013). The formed protective film was dense and tight with high stability in the electrolyte, giving rise to a low interface resistance and a greatly improved cycling CE.

4.2.2 High Coulombic Efficiency and Dendrite Prevention

In Chap. 3, we have described in detail the approaches to enhance CE and prevent Li dendrites in general Li metal-based batteries. In this section, our review will focus on enhancing CE and preventing Li dendrite growth specifically related to Li–S batteries.

4.2.2.1 Effects of Liquid Electrolytes

Organic liquid electrolytes are some of the foremost electrolytes studied in literature on Li–S batteries (Manthiram et al. 2014; Xu et al. 2014a; Chen and Shaw 2014; Scheers et al. 2014; Yin et al. 2013; Bresser et al. 2013; Song et al. 2013b; Zhang 2013b; Barghamadi et al. 2013), probably due to their easy availability, vast variation, excellent wettability on S cathodes and Li anodes, low viscosity to fill in the micropores of the S cathode, relatively high ionic conductivity, low interfacial resistance, reasonable polysulfide solubility, good chemical stability with Li metal anodes, and a good electrochemical window. However, carbonate solvents (both linear and cyclic) are reactive to reduced polysulfide species (Yim et al. 2013; Gao et al. 2011), and sulfone solvents are also incompatible with polysulfide species, according to the poor discharge and charge voltage profiles (Gao et al. 2011); thus, ether-based solvents, including glymes, are the most commonly used organic solvents in Li–S battery studies because of their stability with the Li metal and their compatibility with elemental S cathodes. The Li salts commonly used in Li–S batteries include LiTFSI and $LiSO_3CF_3$ (or LiTf) (Zhang 2013b; Scheers et al. 2014).

Song et al. (2013a) studied the effect of FEC as a co-solvent in electrolytes on the electrochemical performance of the Li metal anode in Li–S batteries. The Li‖Li symmetric cells with the ether-based electrolytes, 1 M $LiPF_6$ in tetra(ethylene glycol) dimethyl ether (Tetraglyme, G4) with and without FEC, showed quite different performances. The Li‖Li cell with the control electrolyte without FEC exploded after 4 h charging, while the symmetric cell with FEC-containing G4-based electrolyte lasted for 70 h without shorting. When the ether-based electrolyte was changed to a carbonate-based electrolyte, i.e., 1 M $LiPF_6$ in EC/EMC (3:7 v/v), the Li‖Li cells with the control electrolyte without FEC or with 5 % FEC showed unstable potential behavior within 5 to 6 Li stripping/deposition cycles and failed at the 7–8 cycles. With an increase in FEC content in the carbonate solvent mixture (EC/EMC/FEC), 60 % FEC could lead to at least 80 stable cycles, or 1600 h of repeated stable Li stripping/deposition processes, while the control electrolyte showed failure at the 10th cycle. At the same time, the 60 % FEC cell showed extremely low polarization voltage (~ 5 mV) after 5 Li stripping/deposition cycles, and the cell impedance for the 60 % FEC cell was only about one-third of that for the control cell. This FEC-derived, SEI-protected Li anode gave stable cycling performance in a Li–S battery with a LiTFSI-TEGDME electrolyte, but it showed decaying cycling stability in the Li–S battery with a $LiPF_6$-TEGDME electrolyte because $LiPF_6$ is not stable with the produced polysulfide species (Zhang 2013b). When the Li electrode was pretreated in a 60 % FEC electrolyte, the preformed LiF-based SEI film could slow the migration of soluble polysulfide species to the Li surface so the surface protective layer was smooth, with no physical fractures and very little S detected. On the contrary, the Li electrode pretreated in a G4-based electrolyte showed significant overcharging during the pre-cycle process in a Li–S battery, indicating the shuttle phenomenon of the soluble polysulfides, which is due to the good compatibility of G4 and long-chain polysulfides (Barchasz et al. 2012).

4.2.2.2 Effects of Lithium Salts

Typically, the salt concentration of an electrolyte is less than 1.2 M due to the considerations of ionic conductivity, viscosity, solubility, and cost. Suo et al. (2013) reported a new class of "solvent-in-salt" electrolytes of LiTFSI/DOL-DME with high salt concentrations up to 7 M. Although the ionic conductivity decreased greatly (from about 15 mS cm^{-1} for 1 M to 0.8 mS cm^{-1} for 7 M) and the viscosity increased significantly (from about 1 cP for 1 M to 72 cP for 7 M) with the increase in salt concentration, the Li^+ transference number increased sharply from about 0.46 for 1–4 M salt concentrations to about 0.73 for 5–7 M. In addition, when this ultrahigh salt concentration (7 M) electrolyte was used in Li–S batteries,

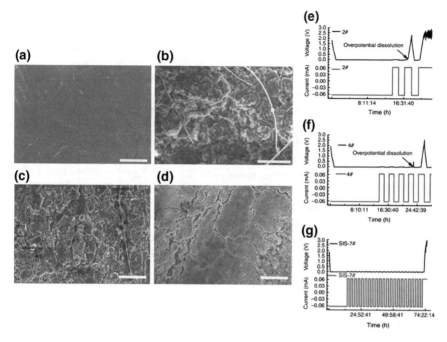

Fig. 4.9 SEM images of Li anodes and Li deposition/stripping profiles: **a** fresh Li metal, **b** Li anode after 278 cycles in 2 M LiTFSI/DOL-DME, **c** Li anode after 183 cycles in 4 M LiTFSI electrolyte, **d** Li anode after 280 cycles in 7 M LiTFSI electrolyte. The *white scale bar* represents 60 μm. **e**, **f**, **g** are for electrolytes with 2 M, 4 M, and 7 M salt concentration, respectively. Reproduced with permission—Copyright 2013, MacMillan Publishers Ltd. (Suo et al. 2013)

it showed very little dissolution of polysulfide species, stable capacity (from an initial 1041 mAh g^{-1} to 770 mAh g^{-1} at the 100th cycle), high capacity retention (74 % for 100 cycles), high CE (from 93.7 % for the first cycle to nearly 100 % after the second cycle), good rate capacity (1229, 988, 864, 744, and 551 mAh g^{-1}-S at current rates of 0.2 C, 0.5 C, 1 C, 2 C, and 3 C, respectively), and good low-temperature performance (about 600, 550, and 350 mAh g^{-1} at −10, −15, and −20 °C, respectively, at 0.1 C rate). However, the lower salt concentration electrolytes resulted in rapid capacity fade, poor capacity retention, and a large variation of CE. More importantly, the concentrated electrolyte provided a good protection of the Li metal anode, showing the lowest roughness and damage levels compared with the Li anodes in low salt concentrations (Fig. 4.9). This demonstrated that high salt concentration electrolytes could significantly diminish the side reactions between the Li metal and the polysulfide species, effectively reduce corrosion of the Li anode, and suppress Li dendrite formation. The reasons include the inhibition of polysulfide dissolution and fewer free solvent molecules in the concentrated electrolytes. It should be noted that low conductivity and high viscosity of concentrated electrolytes may cause significant increase of impedance of Li–S batteries. Recently, a concentrated electrolyte of 5 M LiFSI was also reported to form protective layers on both cathode and anode sides (Kim et al. 2014). Compared to the LiTFSI-based electrolyte, the LiFSI-based electrolyte offered lower solubility of polysulfides, resulting in better performance in Li–S batteries.

Miao et al. (2014) used an electrolyte with dual or binary Li salts, LiFSI, and LiTFSI, dissolved in a mixed ether solvent to address the Li dendrite formation and the low Li CE. When Li was deposited on and stripped from a SS substrate, the efficiency of the stripped charge over the deposited charge reached 91.6 % at the first cycle for the dual-salt electrolyte of 0.5 M LiFSI-0.5 M LiTFSI in DOL-DME (2:1 v/v) and became stable at around 99 % after several cycles, which is much higher than those in the conventional LiPF$_6$/EC-DMC electrolyte for Li-ion batteries and the LiTFSI/DOL-DME electrolyte for Li–S batteries (Fig. 4.10a). The LiFSI-LiTFSI dual-salt electrolyte also resulted in excellent long-term Li deposition/stripping performance for the tested 3075 h, and the overvoltages during the deposition/stripping processes were low (7–15 mV) and stable after more than 300 h cycling (Fig. 4.10d). On the contrary, the two control electrolytes showed a much shorter lifetime of repeated deposition/stripping cycles and the LiPF$_6$/

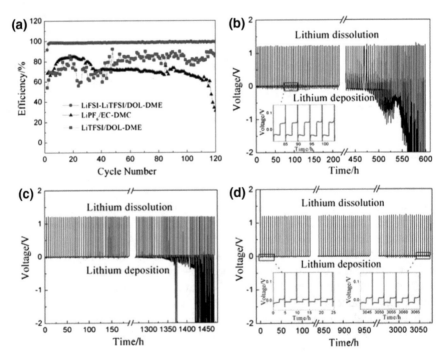

Fig. 4.10 Li deposition/stripping performances in Li|electrolyte|SS cells: **a** CE versus cycle number for three different electrolytes; **b, c, d** galvanostatic voltage–time curves in LiPF$_6$/ EC-DMC, LiTFSI/DOL-DME and LiFSI-LiTFSI/DOL-DME, respectively. Each deposition was conducted at 0.25 mA cm^{-2} for 2.5 h, and the stripping was performed at the same current density with a cutoff voltage of 1.2 V. Reproduced with permission—Copyright 2014, Elsevier (Miao et al. 2014)

EC-DMC electrolyte even had a higher overvoltage, up to about 44 mV (Fig. 4.10b, c, respectively).

4.2.2.3 Effect of Additives

Following successful reports by Mikhaylik (2008) and by Aurbach et al. (2009), $LiNO_3$ has been widely used as a salt additive in electrolytes for Li–S batteries to effectively protect the Li metal anode (Liang et al. 2011b; Zhang 2012a, b, 2013b; Xiong et al. 2011; Scheers et al. 2014; Manthiram et al. 2014). Liang et al. (2011b) reported that the addition of 0.4 M $LiNO_3$ in a 0.5 M $LiCF_3SO_3$/DOL-4G (1:1 v/v) electrolyte could lead to quite a smooth and dense Li surface after 20 cycles. A passive film was found coated on the Li anode surface. Xiong et al. (2012) investigated the components and morphologies of the surface films on the Li anode by using $LiNO_3$ as the sole Li salt in an ether-based electrolyte for Li–S batteries. The strong oxidation of $LiNO_3$ led to its reactions with electrolyte components and the Li metal anode, resulting in a homogeneous surface film containing both inorganic species (such as LiN_xO_y from the direct reduction of $LiNO_3$ by Li metal and Li_2SO_y from the oxidation of sulfur) and organic species (such as ROLi and $ROCO_2Li$), as Aurbach et al. (2009) indicated. When Li metal was immersed in the $LiNO_3$-containing electrolyte, the Li metal surface became more smooth and compact with a longer immersion time (as shown in Fig. 4.11). This homogeneous and compact surface film would enhance the stability of the Li metal anode and improve the cycle life of Li–S batteries.

Besides $LiNO_3$, several other salts and ILs have been reported as electrolyte additives in Li–S batteries to suppress Li dendrite formation. Xiong et al. (2011) studied the effect of LiBOB as an electrolyte additive for electrochemical properties and Li surface morphologies in Li–S batteries. With the increase of LiBOB content in 1 M LiTFSI/DOL-DME (1:1 v/v) from 0 to 10 wt%, the a.c. impedance of the Li anode increased and the discharge capacity of the Li–S battery showed a maximum value at 4 % LiBOB content. However, the cycling stability of Li–S batteries with various LiBOB contents was poor, and nearly no difference could be observed for different LiBOB contents. The authors speculated that the main reasons were attributed to the formation of irreversible Li_2S, structural invalidation of the cathode matrix, and the parasitic reactions between polysulfides and the Li metal. The impact of LiBOB on the cycling performance was small. However, the addition of LiBOB in the electrolyte did form smoother and denser surface morphology on the Li anode, and the roughness of this surface morphology reduced as LiBOB content increased. In contrast, without LiBOB, the Li anode surface showed loosely compacted needle-like dendrites after 50 cycles, causing serious safety concerns. Therefore, the addition of LiBOB as an additive or co-salt could prevent the reactions between the electrolyte and the Li metal.

Wu et al. (2014) investigated the protection of Li anode and improvement of cycling performance for Li–S batteries by using LiDFOB as an additive in the electrolyte of 1 M LiTFSI/DOL-DME. In the control electrolyte without LiDFOB,

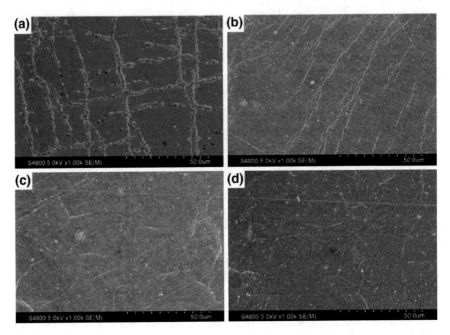

Fig. 4.11 SEM images of the Li metal surface **a** as-received, and immersed in 0.5 M LiNO₃/ DOL-DME (1:1 v/v) for **b** 1 h, **c** 3 h and **d** 5 h. Reproduced with permission—Copyright 2012, Elsevier (Xiong et al. 2011)

the Li–S cell showed poor capacity retention (50 % left after 50 cycles) and fast decay in terms of CE (from 97 % to below 70 % after the first two cycles) due to the severe shuttle phenomenon of polysulfides. With the addition of LiDFOB, both capacity retention and CE were improved; 2 % LiDFOB was the optimal content in this electrolyte, which yielded 70 % capacity retention after 50 cycles and 97 % CE. At 2 % LiDFOB, the a.c. impedance of the Li electrode exhibited its lowest value. The authors attributed this performance improvement to the formation of a good SEI layer on the Li anode surface that suppressed the polysulfide shuttle. The SEM images of the Li electrodes after 50 cycles also showed looser and rougher surface morphologies with a fractured surface for the Li cycled in the control electrolyte without LiDFOB additive (Fig. 4.12a). However, with 2 % LiDFOB the Li surface became relatively smooth except for some small cracks (as indicated by the rectangles in Fig. 4.12b). The XRD, XPS, and density functional theory (DFT) analyses confirmed that LiDFOB promoted the formation of a LiF-rich SEI layer on the Li metal surface, which not only protected and stabilized the Li surface but also blocked the polysulfide shuttle.

Lin et al. studied the effect of P_2S_5 as an additive in liquid electrolytes on the performance of Li–S batteries (Lin et al. 2013b). It was found that P_2S_5 had two

(a)   **(b)**

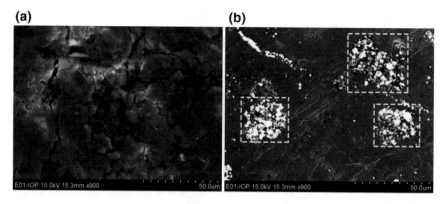

Fig. 4.12 SEM images of Li electrodes after 50 cycles with (a) no LiDFOB and (b) 2 % LiDFOB. Reproduced with permission—Copyright 2014, American Chemical Society (Wu et al. 2014)

functionalities: one was to promote the dissolution of Li_2S in order to alleviate the capacity loss by the precipitation of Li_2S, and the other was to passivate the Li anode surface so as to eliminate the polysulfide shuttle. The passivation layer was about 3–5 μm thick with granular particles less than 100 nm large, which were mainly Li_3PS_4. Due to the superionic conductivity of Li_3PS_4 for Li^+, the thick but highly ionically conductive SEI layer could enhance the Li^+ transportation to and from the Li anode but greatly reduce the polysulfide diffusion and reaction with the Li anode. As a result, the Li metal anode was protected; the CE and cycling performance were improved.

Recently, Zu and Manthiram (2014) reported using copper acetate [Cu $(OOCCH_3)_2$ or $CuAc_2$] as a surface stabilizer for the Li metal anode in Li–S batteries, with an electrolyte of CF_3SO_3Li in DOL/DME with $LiNO_3$. The control cell (without $CuAc_2$) had a sudden capacity decay to 0 mAh g^{-1} at a cycle number close to 100 cycles, likely due to the Li anode degradation. However, with the addition of 0.03 M $CuAc_2$ in the electrolyte, the cell showed a capacity retention of 75 % at the 300th cycle, and a CE of 100 % for most of the cycles; the cell internal resistance was also lower than that of the control cell. The Li anode surfaces showed different morphologies after cycling in the two electrolytes with and without $CuAc_2$. After the first discharge/charge cycle, the Li anode from the control electrolyte contained bulk Li_2S/Li_2S_2 precipitates on the surface and exhibited nonuniformly deposited mossy Li (Fig. 4.13a). The Li anode from the $CuAc_2$-containing electrolyte had a smooth surface and was covered by an S- and Cu-containing passivation film (Fig. 4.13b). After 100 cycles, the Li anode from the control electrolyte showed dendritic morphology and a large amount of S-containing species (Fig. 4.13c), while the Li anode from the $CuAc_2$-containing electrolyte was uniform (Fig. 4.13d). As shown in the cross-sectional energy dispersive spectroscopy (EDS) line scans (Fig. 4.13e, f), S was only detected in the top 20 μm of the surface film on the Li anode cycled in the $CuAc_2$-based electrolyte, indicating the effective blocking of polysulfide penetration, thus preventing the

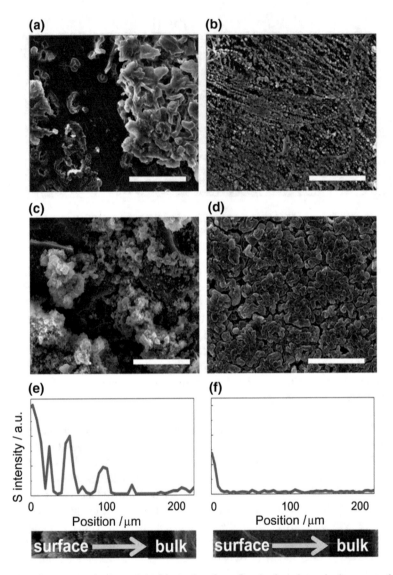

Fig. 4.13 SEM characterizations of the Li metal surface after the first charge in the **a** control cell and **b** experimental cell and after the 100th charge in the **c** control cell and **d** experimental cell. Cross-sectional EDS line scans (for sulfur) of the Li metal after the 100th charge in the **e** control cell and **f** experimental cell. The *scale bars* in **a–d** represent 100 μm. The *scales* in **e** and **f** represent the position in Li metal. The cells were cycled at C/5. Reproduced with permission— Copyright 2014, American Chemical Society (Zu and Manthiram 2014)

formation of Li dendrites. The analyses by EDS, time-of-flight secondary ion mass spectrometry, XRD, and XPS demonstrated the existence of Li_2S, Li_2S_2, CuS, Cu_2S, and other electrolyte decomposition byproducts in the surface film on the Li anode from the $CuAc_2$-containing electrolyte.

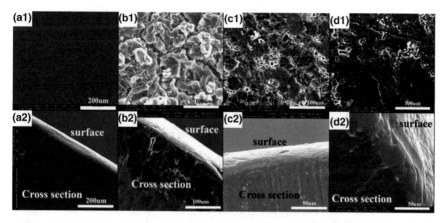

Fig. 4.14 SEM images of the Li metal surface and cross-section of fresh Li metal (*a1* and *a2*) and Li metal cycled in the baseline electrolyte (*b1* and *b2*), electrolyte containing 50 % IL (*c1* and *c2*), and electrolyte containing 75 % IL (*d1* and *d2*) for 100 cycles at 0.2 C. Reproduced with permission—Copyright 2013, The Royal Society of Chemistry (Zheng et al. 2013b)

Zheng et al. also studied the properties and morphologies of the SEI layers formed on the Li metal surface in Li–S batteries by adding the IL 1-butyl-1-methylpyrrolidinium bis(trifluoromethanesulfonyl)-imide ($Pyr_{14}TFSI$) in the electrolyte of 1 M LiTFSI/DME-DOL. It was found that the electrolyte with 75 % $Pyr_{14}TFSI$ in the DME-DOL-$Pyr_{14}TFSI$ mixture gave the Li–S batteries the best cycling stability, highest CE, and highest capacity after 30 cycles among the IL contents from 0 to 100 % in the solvent mixture. The IL-enhanced passivation film on the Li anode surface exhibited very different morphology and chemical composition, effectively protecting the Li metal from continuous attack by soluble polysulfides (see Fig. 4.14) (Zheng et al. 2013b).

4.2.2.4   Effect of Solid Electrolytes

Possible solutions to completely block Li dendrite penetration through the electrolytes include solid-state inorganic electrolytes, which normally are glassy or ceramic, and solid polymer electrolytes. In a recent review paper on Li metal anodes for rechargeable batteries, we have summarized previous achievements in using polymer electrolytes and inorganic solid-state Li-ion conductors to block the penetration of Li dendrites (Xu et al. 2014b). So far, there are several papers reporting the use of such solid-state polymer electrolytes (Marmorstein et al. 2000; Shin et al. 2002; Yu et al. 2004; Hassoun and Scrosati 2010; Liang et al. 2011a) and inorganic electrolytes (Hayashi et al. 2003, 2008; Nagao et al. 2011, 2012a, 2013; Lin et al. 2013a; Hakari et al. 2014) in Li–S related batteries. However, there are no direct reports about using solid-state polymer electrolytes and inorganic electrolytes to suppress Li dendrites in Li–S batteries. We expect that the use of solid-state inorganic electrolytes can prevent Li dendrite growth and penetration in Li–S batteries on the basis of previous reports and knowledge. However, one precaution for using such SSEs is the high interfacial resistance between the solid electrolyte with the S cathode and the Li metal anode. The ability to reduce such interfacial resistances is a big challenge in all solid-state Li–S batteries.

4.2.2.5 Effect of Ex Situ Formed Coating Layers

The protection layers on the Li anode surface are normally formed in situ in liquid electrolytes as discussed above. An alternative approach to protect the Li anode is to pre-form a thin barrier on the Li surface. Kim et al. (2013b) reported the formation of a thin Li–Al alloy layer on the Li surface to protect the Li anode. They laminated a very thin Al foil (0.8 μm) on the Li surface using pressure, and then cured at three different temperatures (25, 60 and 90 °C) for one day. The Li metal with the formed Li–Al thin layer was analyzed and tested in Li–S batteries with a high-concentration electrolyte of 3 M LiTFSI/DOL-DME. The curing temperature affected the electrochemical performance of the Li anode. The Li–Al alloy coating layer cured at 90 °C led to the lowest charge transfer resistance, the lowest polarization voltage, and the best rate capability, which were ascribed to good protection of the Li anode. However, the long-term cycling stability results showed that after 300 cycles, the discharge capacity of the Li–S cell with the Li–Al alloy layer cured at 90 °C was nearly the same as that of a pure Li anode, indicating continuous decay of this alloy protection layer. This is likely related to the high volume change of the Li–Al alloy during de-alloying/re-alloying processes that take place during the charge/discharge cycles.

4.3 Lithium Metal Anodes in Lithium-Air Batteries

Li-air batteries have much higher specific energies than most currently available primary and rechargeable batteries (Hamlen and Atwater 2001; Blurton and Sammells 1979; Arai and Hayashi 2009; Haas and Van Wesemael 2009; Visco et al. 2009; Smedley and Zhang 2009; Zhang 2009; Egashira 2009; Joerissen 2009). A significant amount of work was conducted on Li-air batteries in the 1960s and early 1970s (Gregory 1972; Blurton and Oswin 1972; Oswin 1967); however, efforts in this field decreased considerably in the 1980s because of problems associated with the stability of Li metal anodes and air electrodes, thermal management, and the reversibility of the system. Recent advances in electrode materials and electrolytes, as well as new designs for Li-air batteries, have renewed intensive efforts, especially in the development of Li-air batteries (Abraham and Jiang 1996a, b; Dobley et al. 2006c). In contrast to most other batteries that must carry both anode and cathode inside a storage system, Li-air batteries are unique in that the active cathode material (oxygen) is not stored in the battery. Instead, oxygen can be absorbed from the environment and reduced by catalytic surfaces inside the air electrode. In the typical reaction in a Li-air battery using a nonaqueous electrolyte, the electrolyte will not participate in the reaction, as shown by Eq. (4.1):

$$Li + \frac{1}{2}O_2 \leftrightarrow \frac{1}{2}Li_2O_2 \tag{4.1}$$

Therefore, a very thin layered (i.e., $\sim$0.2–0.3 mm thick) carbon-based air electrode can be used to facilitate electrochemical reactions in metal-air batteries. Figure 4.15 shows the schematic of a Li-air battery based on a nonaqueous electrolyte. One important feature in this battery is that Li^+ is the ion carrier transferred from the anode to the air electrode. As a result, the reaction products are accumulated in the air electrode (cathode) side rather than the anode side of the cell.

Primary aqueous Li-air batteries have been used for decades in applications such as life-vest beacons; the battery is activated by sea water, and the high pH associated with LiOH formation slows Li corrosion sufficiently for operation in this application. However, water corrosion of the anode has hindered development of Li-air batteries for other practical applications (Littauer and Tsai 1977). Several approaches have been used to overcome the water corrosion problem and improve the lifetime of Li-air batteries. Abraham and Jiang (1996b) first reported a Li-air battery based on a nonaqueous electrolyte in 1996. Since then, many groups have conducted extensive work, and have documented the effects of various factors on the performance of Li-air batteries (Abraham and Jiang 1996a, b; Dobley et al. 2006b; Read et al. 2003; Kinoshita 1992; Linden and Reddy 2001; Kuboki et al. 2005; Zhang et al. 2010a; Xiao et al. 2010b; Littauer and Tsai 1977; Ye and Xu

Fig. 4.15 Schematic of reaction processes in a Li-air battery based on a nonaqueous electrolyte. Reproduced with permission —Copyright 2012, American Chemical Society (Shao et al. 2012a)

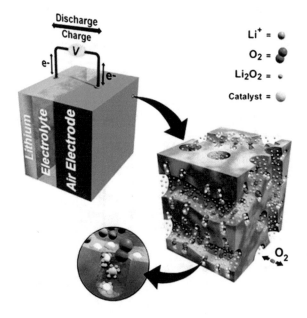

2008; Kumar et al. 2010; Visco et al. 2004a, 2007; Zheng et al. 2008; Beattie et al. 2009; Abraham et al. 1997; Read 2002, 2006; Yang and Xia 2010; Dobley et al. 2006a; Shiga et al. 2008; Williford and Zhang 2009; Xiao et al. 2010a; Xu et al. 2009, 2010; Kowalczk et al. 2007; Wang and Zhou 2010; Shimonishi et al. 2010; Ogasawara et al. 2006; Debart et al. 2007a, b, 2008; Giordani et al. 2010; Laoire et al. 2009). Although a Li anode is relatively stable in a nonaqueous electrolyte, most of these Li-air batteries were investigated in a pure oxygen environment because penetration of moisture from the ambient environment will still corrode a Li electrode. To further extend the lifetime of Li-air batteries, other electrolytes have been investigated in recent years. These electrolytes include ILs (Kuboki et al. 2005; Ye and Xu 2008), SSEs (Kumar et al. 2010), and multilayer electrolytes (Visco et al. 2004a, 2007) consisting of liquid/solid, liquid/solid/solid, and liquid/solid/liquid electrolyte combinations. These multilayer electrolytes (or protected Li anodes [PLEs]) were developed by Visco et al. (2004a, 2007) and have been used successfully in various configurations, including with aqueous electrolytes.

In the battery reported by Abraham and Jiang (1996a, b) and Abraham et al. (1997), the electrolyte consisted of a polymer electrolyte film composed of a mixture of polyacrylonitrile (PAN), EC, PC, and $LiPF_6$ (in a weight ratio of 12:40:40:8, respectively) prepared inside a dry box, heated to 135 °C to obtain a homogeneous solution, and then cooled and rolled into thin membranes. The same electrolyte also was used in a carbon-based air electrode. Figure 4.16 shows the intermittent discharge curve and the open-circuit voltages of a Li|PAN-based polymer electrolyte|oxygen cell at a current density of 0.1 mA cm^{-2} at room temperature in an oxygen atmosphere. The cathode contained Chevron acetylene black carbon. The cell was discharged in 1.5 h increments with an open-circuit stand of about 15 min between discharges. In Fig. 4.16, the open circles represent open-circuit voltage, and the solid line represents load voltage. Abraham and Jiang also presented data suggesting that the main discharge reaction is the reduction of oxygen to form Li_2O_2.

Fig. 4.16 The intermittent discharge curve and the open-circuit voltages of a Li| PAN-based polymer electrolyte|oxygen cell at a current density of 0.1 mA cm^{-2} at room temperature in an oxygen atmosphere. Reproduced with permission—Copyright 1996, Electrochemical Society (Abraham and Jiang 1996b)

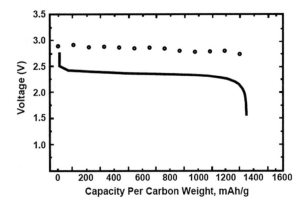

4.3.1 Li-Air Batteries Using Protected Lithium Electrodes

Water corrosion of the Li-metal electrode is one of the main barriers to the practical application of Li-air batteries. Therefore, protection of the Li-metal electrode is a focal point for research in this field. Solid electrolytes, such as LATP glass (Li_{1+x+y} $Al_xTi_{2-x} Si_yP_{3-y}O_{12}$ made by Ohara Inc. of Japan) have good Li-ion conductivity ($\sim 10^{-4}$ S cm^{-1}) and are impermeable to water. This glass is stable in weak acid and alkaline electrolyte. One of the disadvantages of LATP glass is that it is not stable when in contact with Li metal. Visco et al. (2007, 2009) first solved this problem by introducing an interfacial layer (a solid layer such as Cu_3N, LiPON, or nonaqueous electrolyte) between the Li metal and the Ohara glass, thus forming a protected Li electrode (PLE, Fig. 3-26b) (Visco et al. 2007, 2009). Li-air batteries using this PLE can operate in both aqueous and nonaqueous electrolytes. Figure 4. 17 shows the typical discharge curves of Li-air aqueous cells with a protected anode, employing a compliant seal, at various rates: (1) 1.0 mA cm^{-2}; (2) 0. 5 mA cm^{-2}; and (3) 0.2 mA cm^{-2}. The thickness of the Li foil was ~ 2 mm. The end of cell discharge corresponds to Li depletion. Recently, several other groups (Read and Kowalczk et al. (2007), Wang and Zhou (2010), Shimonishi et al. (2010), also used a nonaqueous or polymer electrolyte as the interfacial layer between the Li metal and the LATP glass, thereby forming a triple-electrolyte (nonaqueous electrolyte/LATP/aqueous electrolyte) structure.

4.3.2 Lithium-Air Batteries Using Solid Electrolytes

Kumar et al. (2010) reported on a solid-state Li-air battery. The Li-ion conductive solid electrolyte membrane is based on glass-ceramic (GC) and polymer-ceramic materials. This solid electrolyte is used as the ionic conductive membrane between

Fig. 4.17 Discharge of Li-air aqueous cells with a protected anode employing a compliant seal at the following discharge rates: *1* 1.0 mA cm^{-2}; *2* 0.5 mA cm^{-2}; and *3* 0.2 mA cm^{-2}. The thickness of the Li foil was ~ 2 mm. The end of cell discharge corresponds to Li depletion. Reproduced with permission —Copyright 2009, Elsevier (Visco et al. 2009)

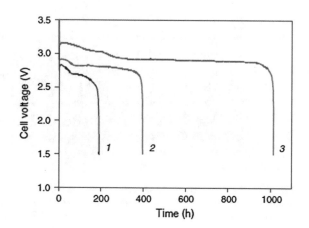

the Li electrode and the air electrode. It also is used in the solid composite air cathode prepared from high-surface area carbon. The cell exhibited excellent thermal stability in the temperature range from 30 to 105 °C temperature range and has been tested for 40 charge-discharge cycles at current densities ranging from 0.05 to 0.25 mA cm^{-2}.

A Li-ion–conducting GC electrolyte was synthesized from the GC membrane prepared from a batch of various oxides. The composition ratio of the membrane was $18.5Li_2O:6.07Al_2O_3:37.05GeO_2:37.05P_2O_5$ (LAGP). A schematic of the cell configuration is shown in Fig. 4.18. A Li anode and a carbon-based cathode are separated by an electrolyte laminate composed of the GC membrane with the composition $5Li_2O:6.07Al_2O_3:37.05GeO_2:37.05P_2O_5$ (LAGP) sandwiched between two polymer composite membranes. The polymer composite adjacent to the anode is composed of LiBETI:PEO (1:8.5) with 1 wt% Li_2O, whereas the polymer composite membrane adjacent to the cathode consists of LiBETI:PEO (1:8.5) with 1 wt% boron nitride. The GC electrolyte used in this work exhibits an ionic conductivity of $\sim 10^{-2}$ S cm^{-1} at 30 °C. The capacity of these solid-state Li-air batteries was found to increase significantly with increasing temperature. This behavior may be attributed to the interfacial resistance in the multilayer structure of Li-air batteries shown in Fig. 4.18.

Although significant progress has been made on primary Li-air batteries, more work needs to be done before rechargeable Li-air batteries can be practically applied. The fundamental reaction mechanisms associated with charge and discharge of a nonaqueous Li-air cell need to be further investigated and understood. In addition, a better bifunctional catalyst needs to be developed to reduce the voltage hysteresis and improve the reversibility of the batteries, and the power rate needs to be improved. Furthermore, in the case of aqueous electrolyte-based Li-air batteries, better protective coatings for the anode are needed, as well as a means of removing the carbon dioxide before it enters the cell. For the nonaqueous Li-air

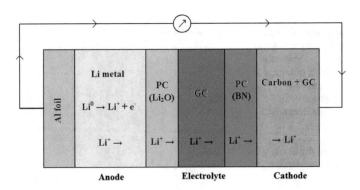

Fig. 4.18 Schematic of the Li-O$_2$ cell and its component materials. The electrolyte laminate is composed of PC/Li$_2$O, GC, and polymer composite (PC/boron nitride) membranes. Reproduced with permission—Copyright American Chemical Society 2010 (Kumar et al. 2010)

design, one of the challenges is to avoid the ingress of water, which may react with Li metal and reduce the lifetime of the batteries.

4.4 Anode-Free Lithium Batteries

A commercial Li-ion battery consists of multiple stacks of anode current collector/anode|separator|cathode/cathode current collector soaked with liquid electrolyte (such as 1 M $LiPF_6$ in carbonate co-solvents). For example, a common Li-ion battery used in commercial electronics has a structure of Cu/graphite|separator|$LiCoO_2$/Al or Cu/graphite|separator|$LiFePO_4$/Al (Jiang et al. 2014; Sun et al. 2012). The operating principle of these Li-ion batteries could be described as follows. During the charging process, the Li ions are extracted out of cathode materials, diffuse through electrolyte soaked in separator, and then intercalate into the anode materials (i.e., graphite). The process is reversed during the discharging process. In this system, each component counts in the total weight and cost of the whole system, although the inactive components including separator, Cu/Al current collectors, and the packaging materials do not contribute to the usable energy.

Ideally, if the Li ions extracted from a cathode can be reversibly deposited onto and stripped from a Cu current collector, then it is possible to assemble a rechargeable Li battery with a structure of Cu|separator|cathode/Al. This battery contains no active anode material as prepared, and can be called an "anode-free Li battery" (AFLB). A schematic illustration of an AFLB concept is presented in Fig. 4.19, where the state-of-the-art Li-ion battery configuration (Fig. 4.19a) is also shown as a reference. In an anode-free battery configuration, all active Li^+ ions are stored inside the cathode electrode (Fig. 4.19b). During the initial charging process, Li ions are extracted from the cathode and diffuse toward the anode current collector, where they are electrodeposited as metallic Li. During the subsequent charging process, Li will be stripped from the anode and intercalated back into the

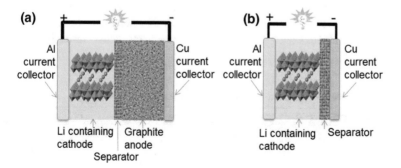

(a) Al current collector — Li containing cathode | Separator | Graphite anode — Cu current collector

(b) Al current collector — Li containing cathode | Separator — Cu current collector

Fig. 4.19 Schematic illustration of **a** state-of-the-art Li-ion battery (i.e., Cu/graphite‖$LiFePO_4$/Al) and **b** an anode-free battery (i.e., Cu‖$LiFePO_4$/Al)

cathode (Ding et al. 2013; Zheng et al. 2014; Lu et al. 2014). Traditionally, this battery structure was considered to be impossible, let alone able to survive for many cycles. In addition to the long-term concern regarding dendritic Li growth that would penetrate through the separator and short the cell (Aurbach et al. 2002; López et al. 2009; Miao et al. 2014), another direct concern with "anode-free" battery structure is that Li cycling CE is very low, usually less than 80 % in most nonaqueous electrolytes. As a consequence, the in situ plated Li supplied from the cathode electrodes will be readily consumed after fewer than ten cycles. In this regard, the key point to realize the anode-free concept is the development of a functional electrolyte in which Li deposition is dendrite free along with high CE and good cycling stability.

The anode-free battery configuration exhibits several advantages. First, the absence of active anode electrode in the as-assembled batteries will reduce the weight and space reserved for the anode. In typical Li-ion batteries, the thickness of the graphite anode is similar to that of the cathode (typically a Li intercalation compound). If the graphite anode can be eliminated in anode-free batteries, the thickness of the active materials (cathode/anode) can be reduced to half of that of Li-ion batteries. Even considering other inactive materials, the energy density (Wh L^{-1}) of the battery can be increased by more than 50 %. Second, it will save energy and cost in anode electrode preparation, including electrode slurry making, slurry coating, and drying. Last but not least, it will operate as a Li metal battery after the initial charge process, providing higher operating voltage and energy density than Li-ion batteries using graphite as the anode.

Although the AFLB has many advantages, tremendous barriers need to be overcome before its practical application. The most critical barrier in these batteries is the very high CE required for an AFLB to have a meaningful lifetime. Table 4.1 shows the cycle life of a battery with different Li cycling CEs. Here we assume that the cathode of the battery has no degradation and the battery fails if its capacity drops to less than 80 % of the original values. At a CE of 99 %, the battery only has a lifetime of 22 cycles. For a rechargeable battery to have a significant lifetime, such as more than 100 cycles, the Li metal anode has to have a CE of at least 99.8 %. At a CE of larger than 99.9 %, an AFLB can have a lifetime of more than a few hundred cycles. At such a high CE, not only the Li metal anode, but all other battery components have to be stable during the electrochemical cycling process. In addition, to evaluate such a high CE, a high precision battery tester is required to

Table 4.1 CE required to retain more than 80 % of the battery capacities

Coulombic efficiency (%)	Cycles to maintain a capacity >80 % of initial value
99	22
99.7	74
99.8	112
99.9	223
99.99	2231

obtain a reliable evaluation of the Li cycling CE. Further development of this technology may lead to practical application of these batteries in the future.

Using solid-state electrolytes such as LiPON has been regarded as a promising solution for anode-free batteries, owing to their good compatibility with Li (Neudecker et al. 2000; Bates et al. 1993, 1995; Lee et al. 2004). However, the intrinsically low conductivity of LiPON and their high cost has limited the application of these batteries (Wang et al. 2015a; Bates et al. 1993, 1995). Therefore, only thin-film electrodes with an areal capacity of ~ 100 μAh cm^{-2} have been reported, limiting their applications in microelectronic devices that do not require high energy or power supply (Neudecker et al. 2000). A liquid organic electrolyte with high conductivity and wettability can be used for the thick electrode in Li metal cells. However, the traditional carbonate-based electrolyte system has poor compatibility with Li metal (due to Li dendrite growth and low CE) and therefore could not be used for anode-free cell design (Woo et al. 2014; Aurbach et al. 2002; Xu et al. 2014b).

Qian et al. (2015b) demonstrated that the highly concentrated electrolytes composed of ether solvents and the LiFSI salt enabled high-rate cycling of a Li metal anode with high CE (up to 99.1 %) without dendrite growth; this was ascribed to the enhanced solvent coordination and improved availability of Li$^+$ ion concentration in the electrolyte. They also demonstrated an anode-free Li cell design (Cu‖LiFePO$_4$) with highly concentrated electrolyte (4 M LiFSI/DME) and delivered high initial discharge capacities close to the nominal capacity of the cathode. The battery retained ~ 60 % of its original capacity after 50 cycles along with an average CE >99 % during cycling. The CE of these batteries can be further improved by adopting a slow charge/fast discharge protocol resulting in an exceptionally high CE of more than 99.8 % (Qian et al. 2016). The in situ formed Li (during charging) is largely stabilized due to minimized reactions between the plated Li and concentrated electrolyte. These works and the insights obtained for the SEI characteristics demonstrate a possible path forward for enabling Li metal as a highly reversible and safe practical battery anode.

References

Abraham KM, Jiang Z (1996a) Solid polymer electrolyte-based oxygen batteries. US 5510209

Abraham KM, Jiang Z (1996b) A polymer electrolyte-based rechargeable lithium/oxygen battery. J Electrochem Soc 143(1):1–5

Abraham KM, Jiang Z, Carroll B (1997) Highly conductive PEO-like polymer electrolytes. Chem Mater 9(9):1978–1988

Agostini M, Aihara Y, Yamada T, Scrosati B, Hassoun J (2013) A lithium–sulfur battery using a solid, glass-type P2S5–Li2S electrolyte. Solid State Ionics 244:48–51. doi:http://dx.doi.org/10.1016/j.ssi.2013.04.024

Arai H, Hayashi M (2009) Secondary batteries-metal-air systems: overview (secondary and primary). In: Garche J, Dyer C, Moseley P, Ogumi Z, Rand D, Scrosati B (eds) Encyclopedia of electrochemical power sources, vol 5. Elsevier, Amsterdam, pp 347–355

Aurbach D, Zinigrad E, Teller H, Dan P (2000) Factors which limit the cycle life of rechargeable lithium (metal) batteries. J Electrochem Soc 147:1274–1279

Aurbach D, Zinigrad E, Cohen Y, Teller H (2002) A short review of failure mechanisms of lithium metal and lithiated graphite anodes in liquid electrolyte solutions. Solid State Ionics 148:405–416

Aurbach D, Pollak E, Elazari R, Salitra G, Kelley CS, Affinito J (2009) On the surface chemical aspects of very high energy density, rechargeable Li–sulfur batteries. J Electrochem Soc 156 (8):A694–A702. doi:10.1149/1.3148721

Autolib' Bluecar carsharing service, Paris, France, by Mariordo (Mario Roberto Durán Ortiz) is licensed under CC BY-SA 3.0 and found on Wikipedia https://en.wikipedia.org/wiki/Bollor% C3%A9_Bluecar#/media/File:Paris_Autolib_06_2012_Bluecar_2907.JPG. Accessed 31 May 2016

Barchasz C, Molton F, Duboc C, Leprêtre J-C, Patoux S, Alloin F (2012) Lithium/sulfur cell discharge mechanism: an original approach for intermediate species identification. Anal Chem 84(9):3973–3980. doi:10.1021/ac2032244

Barchasz C, Lepretre JC, Patoux S, Alloin F (2013a) Revisiting TEGDME/DIOX binary electrolytes for lithium/sulfur batteries: importance of solvation ability and additives. J Electrochem Soc 160(3):A430–A436. doi:10.1149/2.022303jes

Barchasz C, Leprêtre J-C, Patoux S, Alloin F (2013b) Electrochemical properties of ether-based electrolytes for lithium/sulfur rechargeable batteries. Electrochim Acta 89:737–743. doi:10. 1016/j.electacta.2012.11.001

Barghamadi M, Kapoor A, Wen C (2013) A review on Li-S batteries as a high efficiency rechargeable lithium battery. J Electrochem Soc 160(8):A1256–A1263. doi:10.1149/2. 096308jes

Bates JB, Dudney NJ, Gruzalski GR, Zuhr RA, Choudhury A, Luck CF, Robertson JD (1993) Fabrication and characterization of amorphous lithium electrolyte thin-films and rechargeable thin-film batteries. J Power Sources 43(1–3):103–110. doi:10.1016/0378-7753(93)80106-y

Bates JB, Dudney NJ, Lubben DC, Gruzalski GR, Kwak BS, Yu X, Zuhr RA (1995) Thin-film rechargeable lithium batteries. J Power Sources 54(1):58–62. doi:http://dx.doi.org/10.1016/ 0378-7753(94)02040-A

Beattie SD, Manolescu DM, Blair SL (2009) High-capacity lithium-air cathodes. J Electrochem Soc 156(1):A44–A47. doi:10.1149/1.3005989

Blurton KF, Oswin HG (1972) Refuelable batteries. In: Proceedings of the symposium on non-fossil chemical fuels, preprints of papers presented at the 163rd national meeting of the American Chemical Society, vol 2, Boston, Massachusetts, 10–14 April 1972, American Chemical Society, Division of Fuel Chemistry, Washington DC, pp 48–69

Blurton KF, Sammells AF (1979) Metal-air batteries—their status and potential—review. J Power Sources 4(4):263–279

Bresser D, Passerini S, Scrosati B (2013) Recent progress and remaining challenges in sulfur-based lithium secondary batteries—a review. Chem Commun 49(90):10545–10562. doi:10.1039/C3CC46131A

Bruce PG, Freunberger SA, Hardwick LJ, Tarascon J-M (2012) Li-O2 and Li-S batteries with high energy storage. Nat Mater 11(1):19–29

Brückner J, Thieme S, Böttger-Hiller F, Bauer I, Grossmann HT, Strubel P, Althues H, Spange S, Kaskel S (2014) Carbon-based anodes for lithium sulfur full cells with high cycle stability. Adv Funct Mater 24(9):1284–1289. doi:10.1002/adfm.201302169

Bryngelsson H, Stjerndahl M, Gustafsson T, Edström K (2007) How dynamic is the SEI? J Power Sources 174(2):970–975. doi:10.1016/j.jpowsour.2007.06.050

Bucur CB, Muldoon J, Lita A, Schlenoff JB, Ghostine RA, Dietz S, Allred G (2013) Ultrathin tunable ion conducting nanomembranes for encapsulation of sulfur cathodes. Energy Environ Sci 6(11):3286–3290. doi:10.1039/C3EE42739K

Busche MR, Adelhelm P, Sommer H, Schneider H, Leitner K, Janek J (2014) Systematical electrochemical study on the parasitic shuttle-effect in lithium-sulfur-cells at different temperatures and different rates. J Power Sources 259:289–299. doi:10.1016/j.jpowsour.2014. 02.075

Chen L, Shaw LL (2014) Recent advances in lithium–sulfur batteries. J Power Sources 267:770–783. doi:10.1016/j.jpowsour.2014.05.111

Chung S-H, Manthiram A (2014a) Carbonized eggshell membrane as a natural polysulfide reservoir for highly reversible Li-S batteries. Adv Mater 26(9):1360–1365. doi:10.1002/adma. 201304365

Chung S-H, Manthiram A (2014b) A natural carbonized leaf as polysulfide diffusion inhibitor for high-performance lithium-sulfur battery cells. ChemSusChem 7(6):1655–1661. doi:10.1002/cssc.201301287

Chung S-H, Manthiram A (2014c) High-performance Li–S batteries with an ultra-lightweight MWCNT-coated separator. J Phys Chem Lett 5(11):1978–1983. doi:10.1021/jz5006913

Chung SH, Manthiram A (2014d) Bifunctional separator with a light-weight carbon-coating for dynamically and statically stable lithium-sulfur batteries. Adv Funct Mater 24(33):5299–5306. doi:10.1002/adfm.201400845

Debart A, Bao J, Armstrong G, Bruce PG (2007a) Effect of catalyst on the performance of rechargeable lithium/air batteries. ECS Trans 3(27):225–232. doi:10.1149/1.2793594

Debart A, Bao J, Armstrong G, Bruce PG (2007b) An O_2 cathode for rechargeable lithium batteries: the effect of a catalyst. J Power Sources 147(2):1177–1182

Debart A, Paterson AJ, Bao J, Bruce PG (2008) Alpha-MnO_2 nanowires: a catalyst for the O-2 electrode in rechargeable lithium batteries. Angew Chem Int Edit 47(24):4521–4524. doi:10. 1002/anie.200705648

Demir-Cakan R, Morcrette M, Gangulibabu GA, Dedryvere R, Tarascon J-M (2013) Li-S batteries: simple approaches for superior performance. Energy Environ Sci 6(1):176–182. doi:10.1039/C2EE23411D

Ding F, Xu W, Graff GL, Zhang J, Sushko ML, Chen X, Shao Y, Engelhard MH, Nie Z, Xiao J, Liu X, Sushko PV, Liu J, Zhang J-G (2013) Dendrite-free lithium deposition via self-healing electrostatic shield mechanism. J Am Chem Soc 135(11):4450–4456. doi:10.1021/ja312241y

Dobley A, Morein C, Abraham KM (2006a) Cathode optimization for lithium-air batteries. Paper presented at the 208th meeting of the electrochemical society abstracts: energy technology and battery joint general session, 16–21 Oct 2005, Los Angeles, California

Dobley A, Morein C, Roark R, Abraham KM (2006b) Paper presented at the proceedings of the 42nd power sources conference, 12–15 June 2006, Philadelphia, Pennsylvania

Dobley A, Morein C, Roark R, Abraham KM (2006c) Large prototype lithium air batteries. In: Proceedings of the 42nd power sources conference, Philadelphia, Pennsylvania, 12–15 June 2006. U.S. Army Communications-Electronics Command, Fort Monmouth, New Jersey

Egashira M (2009) Secondary batteries-metal-air systems: iron-air (secondary and primary). In: Garche J, Dyer C, Moseley P, Ogumi Z, Rand D, Scrosati B (eds) Encyclopedia of electrochemical power sources, vol 5. Elsevier, Amsterdam, pp 372–375

Elazari R, Salitra G, Gershinsky G, Garsuch A, Panchenko A, Aurbach D (2012) Rechargeable lithiated silicon–sulfur (SLS) battery prototypes. Electrochem Commun 14(1):21–24. doi:10. 1016/j.elecom.2011.10.020

Evers S, Nazar LF (2012) New approaches for high energy density lithium-sulfur battery cathodes. Acc Chem Res 46(5):1135–1143. doi:10.1021/ar3001348

Fu Y, Su Y-S, Manthiram A (2013) Highly reversible lithium/dissolved polysulfide batteries with carbon nanotube electrodes. Angew Chem Int Ed 52(27):6930–6935. doi:10.1002/anie. 201301250

Gao J, Lowe MA, Kiya Y, Abruña HD (2011) Effects of liquid electrolytes on the charge-discharge performance of rechargeable lithium/sulfur batteries: electrochemical and in-situ X-ray absorption spectroscopic studies. J Phys Chem C 115(50):25132–25137. doi:10. 1021/jp207714c

Giordani V, Freunberger SA, Bruce PG, Tarascon JM, Larcher D (2010) H_2O_2 decomposition reaction as selecting tool for catalysts in Li-O-2 cells. Electrochem Solid State Lett 13(12): A180–A183. doi:10.1149/1.3494045

Girishkumar G, McCloskey B, Luntz AC, Swanson S, Wilcke W (2010) Lithium–air battery: promise and challenges. J Phys Chem Lett 1:2193–2203

Gregory DP (1972) Metal-air batteries. Mills and Boon, London

Haas O, Van Wesemael J (2009) Secondary batteries-metal-air systems: zinc-air: electrical recharge. In: Garche J, Dyer C, Moseley P, Ogumi Z, Rand D, Scrosati B (eds) Encyclopedia of electrochemical power sources, vol 5. Elsevier, Amsterdam, pp 384–392

Hakari T, Nagao M, Hayashi A, Tatsumisago M (2014) Preparation of composite electrode with Li2S–P2S5 glasses as active materials for all-solid-state lithium secondary batteries. Solid State Ionics 262:147–150. doi:10.1016/j.ssi.2013.09.023

Hamlen RP, Atwater TB (2001) Metal/air batteries. In: Linden D, Reddy T (eds) Handbook of batteries, 3rd edn. McGraw Hill, New York, pp 38.31–38.53

Hassoun J, Scrosati B (2010) A high-performance polymer tin sulfur lithium ion battery. Angew Chem Int Ed 49(13):2371–2374. doi:10.1002/anie.200907324

Hayashi A, Ohtomo T, Mizuno F, Tadanaga K, Tatsumisago M (2003) All-solid-state Li/S batteries with highly conductive glass–ceramic electrolytes. Electrochem Commun 5(8):701–705. doi:10.1016/S1388-2481(03)00167-X

Hayashi A, Ohtsubo R, Ohtomo T, Mizuno F, Tatsumisago M (2008) All-solid-state rechargeable lithium batteries with Li2S as a positive electrode material. J Power Sources 183(1):422–426. doi:10.1016/j.jpowsour.2008.05.031

Heine J, Krüger S, Hartnig C, Wietelmann U, Winter M, Bieker P (2014) Coated lithium powder (CLiP) electrodes for lithium-metal batteries. Adv Energy Mater 4(5). doi:10.1002/aenm. 201300815

Huang J-Q, Zhang Q, Zhang S-M, Liu X-F, Zhu W, Qian W-Z, Wei F (2013) Aligned sulfur-coated carbon nanotubes with a polyethylene glycol barrier at one end for use as a high efficiency sulfur cathode. Carbon 58:99–106. doi:10.1016/j.carbon.2013.02.037

Huang C, Xiao J, Shao Y, Zheng J, Bennett WD, Lu D, Saraf LV, Engelhard M, Ji L, Zhang J, Li X, Graff GL, Liu J (2014a) Manipulating surface reactions in lithium-sulphur batteries using hybrid anode structures. Nat Commun 5:3015. doi:10.1038/ncomms4015

Huang J-Q, Zhang Q, Peng H-J, Liu X-Y, Qian W-Z, Wei F (2014b) Ionic shield for polysulfides towards highly-stable lithium-sulfur batteries. Energy Environ Sci 7(1):347–353. doi:10.1039/C3EE42223B

Jeddi K, Zhao Y, Zhang Y, Konarov A, Chen P (2013) Fabrication and characterization of an effective polymer nanocomposite electrolyte membrane for high performance lithium/sulfur batteries. J Electrochem Soc 160(8):A1052–A1060. doi:10.1149/2.010308jes

Jeddi K, Sarikhani K, Qazvini NT, Chen P (2014) Stabilizing lithium/sulfur batteries by a composite polymer electrolyte containing mesoporous silica particles. J Power Sources 245:656–662. doi:10.1016/j.jpowsour.2013.06.147

Jeong T-G, Moon YH, Chun H-H, Kim HS, Cho BW, Kim Y-T (2013) Free standing acetylene black mesh to capture dissolved polysulfide in lithium sulfur batteries. Chem Commun 49 (94):11107–11109. doi:10.1039/C3CC46358C

Ji XL, Lee KT, Nazar LF (2009) Nature Mater 8:500–506

Jiang J, Shi W, Zheng J, Zuo P, Xiao J, Chen X, Xu W, Zhang J-G (2014) Optimized operating range for large-format LiFePO4/graphite batteries. J Electrochem Soc 161(3):A336–A341. doi:10.1149/2.052403jes

Jin Z, Xie K, Hong X (2013a) Electrochemical performance of lithium/sulfur batteries using perfluorinated ionomer electrolyte with lithium sulfonyl dicyanomethide functional groups as functional separator. RSC Adv 3(23):8889–8898. doi:10.1039/C3RA41517A

Jin Z, Xie K, Hong X, Hu Z (2013b) Capacity fading mechanism in lithium sulfur cells using poly (ethylene glycol)-borate ester as plasticizer for polymer electrolytes. J Power Sources 242:478–485. doi:10.1016/j.jpowsour.2013.05.086

Joerissen L (2009) Secondary batteries-metal-air systems: bifunctional oxygen electrodes. In: Garche J, Dyer C, Moseley P, Ogumi Z, Rand D, Scrosati B (eds) Encyclopedia of electrochemical power sources, vol 5. Elsevier, Amsterdam, pp 356–371

Kim H, Lee JT, Yushin G (2013a) High temperature stabilization of lithium–sulfur cells with carbon nanotube current collector. J Power Sources 226:256–265. doi:10.1016/j.jpowsour. 2012.10.028

Kim H, Lee JT, Lee D-C, Oschatz M, Cho WI, Kaskel S, Yushin G (2013b) Enhancing performance of Li–S cells using a Li–Al alloy anode coating. Electrochem Commun 36:38–41. doi:10.1016/j.elecom.2013.09.002

Kim H, Wu F, Lee JT, Nitta N, Lin H-T, Oschatz M, Cho WI, Kaskel S, Borodin O, Yushin G (2014) In situ formation of protective coatings on sulfur cathodes in lithium batteries with LiFSI-based organic electrolytes. Adv Energy Mater. doi:10.1002/aenm.201401792

Kinoshita K (1992) Electrochemical oxygen technology. The electochemical society series. John Wiley & Sons, New York

Kinoshita S, Okuda K, Machida N, Shigematsu T (2014) Additive effect of ionic liquids on the electrochemical property of a sulfur composite electrode for all-solid-state lithium–sulfur battery. J Power Sources 269:727–734. doi:10.1016/j.jpowsour.2014.07.055

Kowalczk I, Read J, Salomon M (2007) Li-air batteries: a classic example of limitations owing to solubilities. Pure Appl Chem 79(5):851–860. doi:10.1351/pac200779050851

Kuboki T, Okuyama T, Ohsaki T, Takami N (2005) Lithium-air batteries using hydrophobic room temperature ionic liquid electrolyte. J Power Sources 146(1–2):766–769. doi:10.1016/j. jpowsour.2005.03.082

Kulisch J, Sommer H, Brezesinski T, Janek J (2014) Simple cathode design for Li-S batteries: cell performance and mechanistic insights by in operando X-ray diffraction. Phys Chem Chem Phys 16(35):18765–18771. doi:10.1039/C4CP02220C

Kumar B, Kumar J, Leese R, Fellner JP, Rodrigues SJ, Abraham KM (2010) A solid-state, rechargeable, long cycle life lithium-air battery. J Electrochem Soc 157(1):A50–A54. doi:10. 1149/1.3256129

Laoire CO, Mukerjee S, Abraham KM, Plichta EJ, Hendrickson MA (2009) Elucidating the mechanism of oxygen reduction for lithium-air battery applications. J Phys Chem C 113 (46):20127–20134. doi:10.1021/jp908090s

Lee YM, Choi N-S, Park JH, Park J-K (2003) Electrochemical performance of lithium/sulfur batteries with protected Li anodes. J Power Sources 119–121:964–972. doi:10.1016/s0378-7753(03)00300-8

Lee S-H, Tracy E, Liu P (2004) Buried anode lithium thin film battery and process for forming the same. USA Patent 6,805,999, 19 October 2004

Lee J-S, Kim ST, Cao R, Choi N-S, Liu M, Lee KT, Cho J (2011) Adv Energy Mater 1:34–50

Li W, Hicks-Garner J, Wang J, Liu J, Gross AF, Sherman E, Graetz J, Vajo JJ, Liu P (2014) V2O5 polysulfide anion barrier for long-lived Li-S batteries. Chem Mater 26(11):3404–3410

Liang X, Wen Z, Liu Y, Zhang H, Huang L, Jin J (2011a) Highly dispersed sulfur in ordered mesoporous carbon sphere as a composite cathode for rechargeable polymer Li/S battery. J Power Sources 196(7):3655–3658. doi:10.1016/j.jpowsour.2010.12.052

Liang X, Wen Z, Liu Y, Wu M, Jin J, Zhang H, Wu X (2011b) Improved cycling performances of lithium sulfur batteries with LiNO3-modified electrolyte. J Power Sources 196(22):9839–9843. doi:10.1016/j.jpowsour.2011.08.027

Lin Z, Liu Z, Dudney NJ, Liang C (2013a) Lithium superionic sulfide cathode for all-solid lithium-sulfur batteries. ACS Nano 7(3):2829–2833. doi:10.1021/nn400391h

Lin Z, Liu Z, Fu W, Dudney NJ, Liang C (2013b) Phosphorous pentasulfide as a novel additive for high-performance lithium-sulfur batteries. Adv Funct Mater 23(8):1064–1069. doi:10.1002/ adfm.201200696

Linden D, Reddy T (2001) Handbook of batteries, 3rd edn. McGraw-Hill, New York

Littauer EL, Tsai KC (1977) Corrosion of lithium in alkaline-solution. J Electrochem Soc 124 (6):850–855

Liu C, Ma X, Xu F, Zheng L, Zhang H, Feng W, Huang X, Armand M, Nie J, Chen H, Zhou Z (2014a) Ionic liquid electrolyte of lithium bis(fluorosulfonyl)imide/N-Methyl-N-propylpiperidinium bis(fluorosulfonyl)imide for Li/natural graphite cells: effect of concentration of lithium salt on the physicochemical and electrochemical properties. Electrochim Acta 149:370–385. doi:10.1016/j.electacta.2014.10.048

Liu Z, Zhang X-H, Lee C-S (2014b) A stable high performance Li-S battery with a polysulfide ion blocking layer. J Mater Chem A 2(16):5602–5605. doi:10.1039/C4TA00015C

López CM, Vaughey JT, Dees DW (2009) Morphological transitions on lithium metal anodes. J Electrochem Soc 156(9):A726–A729. doi:10.1149/1.3158548

Lu Y, Tu Z, Archer LA (2014) Stable lithium electrodeposition in liquid and nanoporous solid electrolytes. Nat Mater 13(10):961–969. doi:10.1038/nmat4041

Lv D, Shao Y, Lozano T, Bennett WD, Graff GL, Polzin B, Zhang J-G, Engelhard MH, Saenz NT, Henderson WA, Bhattacharya P, Liu J, Xiao J (2015) Failure mechanism for fast-charged lithium metal batteries with liquid electrolytes. Adv Energy Mater 5(3):1400993. doi:10.1002/aenm.201400993

Ma G, Wen Z, Jin J, Wu M, Wu X, Zhang J (2014) Enhanced cycle performance of Li–S battery with a polypyrrole functional interlayer. J Power Sources 267:542–546. doi:10.1016/j.jpowsour.2014.05.057

Manthiram A, Fu Y, Su Y-S (2012) Challenges and prospects of lithium-sulfur batteries. Acc Chem Res 46(5):1125–1134. doi:10.1021/ar300179v

Manthiram A, Fu Y, Chung SH, Zu C, Su YS (2014) Rechargeable lithium-sulfur batteries. Chem Rev 114(23):11751–11787. doi:10.1021/cr500062v

Marmorstein D, Yu TH, Striebel KA, McLarnon FR, Hou J, Cairns EJ (2000) Electrochemical performance of lithium/sulfur cells with three different polymer electrolytes. J Power Sources 89(2):219–226. doi:10.1016/S0378-7753(00)00432-8

Miao R, Yang J, Feng X, Jia H, Wang J, Nuli Y (2014) Novel dual-salts electrolyte solution for dendrite-free lithium-metal based rechargeable batteries with high cycle reversibility. J Power Sources 271:291–297. doi:10.1016/j.jpowsour.2014.08.011

Mikhaylik YV (2008) Electrolytes for lithium sulfur cells. USA Patent 7,354,680, 8 April 2008

Munichandraiah N, Scanlon LG, Marsh RA (1998) Surface films of lithium: an overview of electrochemical studies. J Power Sources 72:203–210

Nagao M, Hayashi A, Tatsumisago M (2011) Sulfur–carbon composite electrode for all-solid-state Li/S battery with Li2S–P2S5 solid electrolyte. Electrochim Acta 56(17):6055–6059. doi:10.1016/j.electacta.2011.04.084

Nagao M, Hayashi A, Tatsumisago M (2012a) High-capacity Li2S-nanocarbon composite electrode for all-solid-state rechargeable lithium batteries. J Mater Chem 22(19):10015–10020. doi:10.1039/C2JM16802B

Nagao M, Hayashi A, Tatsumisago M (2012b) Fabrication of favorable interface between sulfide solid electrolyte and Li metal electrode for bulk-type solid-state Li/S battery. Electrochem Commun 22:177–180. doi:10.1016/j.elecom.2012.06.015

Nagao M, Imade Y, Narisawa H, Kobayashi T, Watanabe R, Yokoi T, Tatsumi T, Kanno R (2013) All-solid-state Li–sulfur batteries with mesoporous electrode and thio-LISICON solid electrolyte. J Power Sources 222:237–242. doi:10.1016/j.jpowsour.2012.08.041

Nazar LF, Cuisinier M, Pang Q (2014) Lithium-sulfur batteries. MRS Bull 39(05):436–442. doi:10.1557/mrs.2014.86

Neudecker BJ, Dudney NJ, Bates JB (2000) "Lithium-free" thin-film battery with in situ plated Li anode. J Electrochem Soc 147(2):517–523

Ogasawara T, Debart A, Holzapfel M, Novak P, Bruce PG (2006) Rechargeable Li$_2$O$_2$ electrode for lithium batteries. J Am Chem Soc 128(4):1390–1393. doi:10.1021/ja056811q

Oh S, Yoon W (2014) Effect of polypyrrole coating on Li powder anode for lithium-sulfur secondary batteries. Int J Precis Eng Manuf 15(7):1453–1457. doi:10.1007/s12541-014-0490-y

Oh SJ, Lee JK, Yoon WY (2014) Preventing the dissolution of lithium polysulfides in lithium-sulfur cells by using Nafion-coated cathodes. ChemSusChem 7(9):2562–2566. doi:10.1002/cssc.201402318

Oswin HG (1967) Performance forecast of selected static energy conversion devices. In: Sherman GW, Devol L (eds) Proceedings of the 29th meeting of AGARD propulsion and energetics panel, Liege, Belgium, 12–16 June 1967. Air Force Aero Proplusion Laboratory, Wright-Patterson Air Force Base, Ohio, p 397

Park J-W, Ueno K, Tachikawa N, Dokko K, Watanabe M (2013) Ionic liquid electrolytes for lithium-sulfur batteries. J Phys Chem C 117(40):20531–20541. doi:10.1021/jp408037e

Qian J, Henderson WA, Xu W, Bhattacharya P, Engelhard M, Borodin O, Zhang JG (2015) High rate and stable cycling of lithium metal anode. Nat Commun 6:6362. doi:10.1038/ncomms7362

Qian JF, Adams BD, Zheng JM, Xu W, Henderson WA, Wang J, Bowden ME, Xu SC, Hu JZ, Zhang JZ (2016) Anode-free rechargeable lithium metal batteries. Adv Funct Mater 2016. doi:10.1002/adfm.201602353

Ratnakumar BV, Smart MC, Surampudi S (2001) Effects of SEI on the kinetics of lithium intercalation. J Power Sources 97–98:137–139. doi:10.1016/S0378-7753(01)00682-6

Read J (2002) Characterization of the lithium/oxygen organic electrolyte battery. J Electrochem Soc 149(9):A1190–A1195. doi:10.1149/1.1498256

Read J (2006) Ether-based electrolytes for the lithium/oxygen organic electrolyte battery. J Electrochem Soc 153(1):A96–A100. doi:10.1149/1.2131827

Read J, Mutolo K, Ervin M, Behl W, Wolfenstine J, Driedger A, Foster D (2003) Oxygen transport properties of organic electrolytes and performance of lithium/oxygen battery. J Electrochem Soc 150(10):A1351–A1356. doi:10.1149/1.1606454

Scheers J, Fantini S, Johansson P (2014) A review of electrolytes for lithium-sulphur batteries. J Power Sources 255:204–218. doi:10.1016/j.jpowsour.2014.01.023

Shao Y, Park S, Xiao J, Zhang J-G, Wang Y, Liu AJ (2012a) Electrocatalysts for nonaqueous lithium-air batteries: status. Challenges, and Perspective. ACS Catalysis 845

Shiga T, Nakano H, Imagawa H (2008) Non-aqueous air battery and catalyst therefor. US Patent Application 2008/0299456

Shimonishi Y, Zhang T, Johnson P, Imanishi N, Hirano A, Takeda Y, Yamamoto O, Sammes N (2010) A study on lithium/air secondary batteries-Stability of NASICON-type glass ceramics in acid solutions. J Power Sources 195(18):6187–6191. doi:10.1016/j.jpowsour.2009.11.023

Shin JH, Kim KW, Ahn HJ, Ahn JH (2002) Electrochemical properties and interfacial stability of (PEO)10LiCF3SO3–TinO2n − 1 composite polymer electrolytes for lithium/sulfur battery. Mater Sci Eng, B 95(2):148–156. doi:10.1016/S0921-5107(02)00226-X

Smedley S, Zhang XG (2009) Secondary batteries-metal-air systems: zinc-air: hydraulic recharge. In: Garche J, Dyer C, Moseley P, Ogumi Z, Rand D, Scrosati B (eds) Encyclopedia of electrochemical power sources, vol 5. Elsevier, Amsterdam, pp 393–403

Song JH, Yeon JT, Jang JY, Han JG, Lee SM, Choi NS (2013a) Effect of fluoroethylene carbonate on electrochemical performances of lithium electrodes and lithium-sulfur batteries. J Electrochem Soc 160(6):A873–A881. doi:10.1149/2.101306jes

Song MK, Cairns EJ, Zhang Y (2013b) Lithium/sulfur batteries with high specific energy: old challenges and new opportunities. Nanoscale 5(6):2186–2204. doi:10.1039/c2nr33044j

Su Y-S, Manthiram A (2012) Lithium–sulphur batteries with a microporous carbon paper as a bifunctional interlayer. Nat Commun 3:1166

Sun Y-K, Chen Z, Noh H-J, Lee D-J, Jung H-G, Ren Y, Wang S, Yoon CS, Myung S-T, Amine K (2012) Nanostructured high-energy cathode materials for advanced lithium batteries. Nat Mater 11(11):942–947. doi:10.1038/nmat3435

Suo L, Hu YS, Li H, Armand M, Chen L (2013) A new class of solvent-in-salt electrolyte for high-energy rechargeable metallic lithium batteries. Nat Commun 4:1481. doi:10.1038/ncomms2513

Tang Q, Shan Z, Wang L, Qin X, Zhu K, Tian J, Liu X (2014) Nafion coated sulfur–carbon electrode for high performance lithium–sulfur batteries. J Power Sources 246:253–259. doi:10.1016/j.jpowsour.2013.07.076

Tobishima S, Sakurai Y, Yamaki J (1996) Safety characteristics of rechargeable lithium metal cells. In: 8th international meeting on lithium batteries, p 362. doi:10.1016/S0378-7753(96)02584-0

Unemoto A, Ogawa H, Gambe Y, Honma I (2014) Development of lithium-sulfur batteries using room temperature ionic liquid-based quasi-solid-state electrolytes. Electrochim Acta 125:386–394. doi:10.1016/j.electacta.2014.01.105

Vaughey JT, Liu G, Zhang J-G (2014) Stabilizing the surface of lithium metal. MRS Bull 39 (05):429–435. doi:10.1557/mrs.2014.88

Visco SJ, Nimon E, De Jonghe LC, Katz B, Chu MY (2004a) Lithium fuel cells. Paper presented at the proceedings of the 12th international meeting on lithium batteries, 27 June–2 July 2004, Nara, Japan

Visco SJ, Katz B, Nimon YS, De Jonghe LC (2007) Protected active metal electrode and battery cell structures with non-aqueous interlayer architecture. US 7282295

Visco SJ, Nimon E, De Jonghe C (2009) Secondary batteries-metal-air systems: lithium-air. In: Garche J, Dyer C, Moseley P, Ogumi Z, Rand D, Scrosati B (eds) Encyclopedia of electrochemical power sources, vol 5. Elsevier, Amsterdam, pp 376–383

Wang YG, Zhou HS (2010) A lithium-air battery with a potential to continuously reduce O-2 from air for delivering energy. J Power Sources 195(1):358–361. doi:10.1016/j.jpowsour.2009.06.109

Wang X, Wang Z, Chen L (2013) Reduced graphene oxide film as a shuttle-inhibiting interlayer in a lithium–sulfur battery. J Power Sources 242:65–69. doi:10.1016/j.jpowsour.2013.05.063

Wang J, Lin F, Jia H, Yang J, Monroe CW, NuLi Y (2014) Towards a safe lithium-sulfur battery with a flame-inhibiting electrolyte and a sulfur-based composite cathode. Angew Chem Int Ed 53(38):10099–10104. doi:10.1002/anie.201405157

Wang D, Zhong G, Li Y, Gong Z, McDonald MJ, Mi J-X, Fu R, Shi Z, Yang Y (2015) Enhanced ionic conductivity of $Li_{3.5}Si_{0.5}P_{0.5}O_4$ with addition of lithium borate. Solid State Ionics 283:109–114. doi:10.1016/j.ssi.2015.10.009

Whittingham MS (2004) Lithium batteries and cathode materials. Chem Rev 104(10):4271–4302. doi:10.1021/cr020731c

Whittingham MS (2012) History, evolution, and future status of energy storage. In: Proceedings of IEE 100 (special centennial issue), 1518–1534. doi:10.1109/JPROC.2012.2190170

Williford RE, Zhang JG (2009) Air electrode design for sustained high power operation of Li/air batteries. J Power Sources 194(2):1164–1170. doi:10.1016/j.jpowsour.2009.06.005

Woo J-J, Maroni VA, Liu G, Vaughey JT, Gosztola DJ, Amine K, Zhang Z (2014) Symmetrical impedance study on inactivation induced degradation of lithium electrodes for batteries beyond lithium-ion. J Electrochem Soc 161(5):A827–A830. doi:10.1149/2.089405jes

Wu M, Wen Z, Jin J, Cui Y (2013) Effects of combinatorial $AlCl_3$ and pyrrole on the SEI formation and electrochemical performance of Li electrode. Electrochim Acta 103:199–205. doi:10.1016/j.electacta.2013.03.181

Wu F, Qian J, Chen R, Lu J, Li L, Wu H, Chen J, Zhao T, Ye Y, Amine K (2014) An effective approach to protect lithium anode and improve cycle performance for Li-S batteries. ACS Appl Mater Interfaces 6(17):15542–15549. doi:10.1021/am504345s

Xiao J, Xu W, Wang DY, Zhang JG (2010a) Hybrid air-electrode for Li/air batteries. J Electrochem Soc 157(3):A294–A297. doi:10.1149/1.3280281

Xiao J, Wang DH, Xu W, Wang DY, Williford RE, Liu J, Zhang JG (2010b) Optimization of air electrode for Li/air batteries. J Electrochem Soc 157(4):A487–A492. doi:10.1149/1.3314375

Xiong S, Kai X, Hong X, Diao Y (2011) Effect of LiBOB as additive on electrochemical properties of lithium-sulfur batteries. Ionics 18(3):249–254. doi:10.1007/s11581-011-0628-1

Xiong S, Xie K, Diao Y, Hong X (2012) Properties of surface film on lithium anode with $LiNO_3$ as lithium salt in electrolyte solution for lithium-sulfur batteries. Electrochim Acta 83:78–86. doi:10.1016/j.electacta.2012.07.118

Xiong S, Xie K, Diao Y, Hong X (2013) On the role of polysulfides for a stable solid electrolyte interphase on the lithium anode cycled in lithium-sulfur batteries. J Power Sources 236:181–187. doi:10.1016/j.jpowsour.2013.02.072

Xiong S, Xie K, Diao Y, Hong X (2014a) Characterization of the solid electrolyte interphase on lithium anode for preventing the shuttle mechanism in lithium-sulfur batteries. J Power Sources 246:840–845. doi:10.1016/j.jpowsour.2013.08.041

Xiong S, Xie K, Blomberg E, Jacobsson P, Matic A (2014b) Analysis of the solid electrolyte interphase formed with an ionic liquid electrolyte for lithium-sulfur batteries. J Power Sources 252:150–155. doi:10.1016/j.jpowsour.2013.11.119

Xiong S, Diao Y, Hong X, Chen Y, Xie K (2014c) Characterization of solid electrolyte interphase on lithium electrodes cycled in ether-based electrolytes for lithium batteries. J Electroanal Chem 719:122–126. doi:10.1016/j.jelechem.2014.02.014

Xu W, Xiao J, Zhang J, Wang DY, Zhang JG (2009) Optimization of nonaqueous electrolytes for primary lithium/air batteries operated in ambient environment. J Electrochem Soc 156(10):A773–A779. doi:10.1149/1.3168564

Xu W, Xiao J, Wang DY, Zhang J, Zhang JG (2010) Effects of nonaqueous electrolytes on the performance of lithium/air batteries. J Electrochem Soc 157(2):A219–A224. doi:10.1149/1.3269928

Xu G, Ding B, Pan J, Nie P, Shen L, Zhang X-W (2014a) High performance lithium-sulfur batteries: advances and challenges. J Mater Chem A 2(32):12662–12676. doi:10.1039/c4ta02097a

Xu W, Wang J, Ding F, Chen X, Nasybulin E, Zhang Y, Zhang J-G (2014b) Lithium metal anodes for rechargeable batteries. Energy Environ Sci 7(2):513–537. doi:10.1039/c3ee40795k

Yamin H, Peled E (1983) Electrochemistry of a nonaqueous lithium/sulfur cell. J Power Sources 9:281–287

Yang XH, Xia YY (2010) The effect of oxygen pressures on the electrochemical profile of lithium/oxygen battery. J Solid State Electrochem 14(1):109–114. doi:10.1007/s10008-009-0791-8

Yang Y, McDowell MT, Jackson A, Cha JJ, Hong SS, Cui Y (2010) New nanostructured Li₂S/silicon rechargeable battery with high specific energy. Nano Lett 10(4):1486–1491. doi:10.1021/nl100504q

Yang Y, Zheng G, Cui Y (2013a) Nanostructured sulfur cathodes. Chem Soc Rev 42(7):3018–3032. doi:10.1039/C2CS35256G

Yang Y, Zheng G, Cui Y (2013b) A membrane-free lithium/polysulfide semi-liquid battery for large-scale energy storage. Energy Environ Sci 6(5):1552–1558. doi:10.1039/C3EE00072A

Yao H, Yan K, Li W, Zheng G, Kong D, Seh ZW, Narasimhan VK, Liang Z, Cui Y (2014) Improved lithium-sulfur batteries with a conductive coating on the separator to prevent the accumulation of inactive S-related species at the cathode-separator interface. Energy Environ Sci. doi:10.1039/C4EE01377H

Ye H, Xu JJ (2008) Polymer electrolytes based on ionic liquids and their application to solid-state thin-film Li-oxygen batteries. ECS Trans 3(42):73–81. doi:10.1149/1.2838194

Yim T, Park M-S, Yu J-S, Kim KJ, Im KY, Kim J-H, Jeong G, Jo YN, Woo S-G, Kang KS, Lee I, Kim Y-J (2013) Effect of chemical reactivity of polysulfide toward carbonate-based electrolyte on the electrochemical performance of Li–S batteries. Electrochim Acta 107:454–460. doi:10.1016/j.electacta.2013.06.039

Yin Y-X, Xin S, Guo Y-G, Wan L-J (2013) Lithium-sulfur batteries: electrochemistry, materials, and prospects. Angew Chem Int Ed 52(50):13186–13200. doi:10.1002/anie.201304762

Younesi R, Hahlin M, Roberts M, Edström K (2013) The SEI layer formed on lithium metal in the presence of oxygen: a seldom considered component in the development of the Li–O₂ battery. J Power Sources 225:40–45. doi:10.1016/j.jpowsour.2012.10.011

Yu X, Xie J, Yang J, Wang K (2004) All solid-state rechargeable lithium cells based on nano-sulfur composite cathodes. J Power Sources 132(1–2):181–186. doi:10.1016/j.jpowsour.2004.01.034

Yu M, Yuan W, Li C, Hong J-D, Shi G (2014) Performance enhancement of a graphene-sulfur composite as a lithium-sulfur battery electrode by coating with an ultrathin Al₂O₃ film via atomic layer deposition. J Mater Chem A 2(20):7360–7366. doi:10.1039/C4TA00234B

Zaghib K (2012) Lithium metal for rechargeable polymer and metal-air batteries: challenges and opportunities

Zhang XG (2009) Zinc electrodes: overview. In: Garche J, Dyer C, Moseley P, Ogumi Z, Rand D, Scrosati B (eds) Encyclopeida of electrochemical power sources, vol 5. Elsevier, Amsterdam, pp 454–468

Zhang SS (2012a) Role of LiNO$_3$ in rechargeable lithium/sulfur battery. Electrochim Acta 70:344–348. doi:10.1016/j.electacta.2012.03.081

Zhang SS (2012b) Effect of discharge cutoff voltage on reversibility of lithium/sulfur batteries with LiNO$_3$-contained electrolyte. J Electrochem Soc 159(7):A920–A923. doi:10.1149/2.002207jes

Zhang SS (2013a) Liquid electrolyte lithium/sulfur battery: fundamental chemistry, problems, and solutions. J Power Sources 231:153–162. doi:10.1016/j.jpowsour.2012.12.102

Zhang SS (2013b) A concept for making poly(ethylene oxide) based composite gel polymer electrolyte lithium/sulfur battery. J Electrochem Soc 160(9):A1421–A1424. doi:10.1149/2.058309jes

Zhang SS, Read JA (2012) A new direction for the performance improvement of rechargeable lithium/sulfur batteries. J Power Sources 200:77–82. doi:10.1016/j.jpowsour.2011.10.076

Zhang SS, Tran DT (2013) How a gel polymer electrolyte affects performance of lithium/sulfur batteries. Electrochim Acta 114:296–302. doi:10.1016/j.electacta.2013.10.069

Zhang JG, Wang DY, Xu W, Xiao J, Williford RE (2010) Ambient operation of Li/Air batteries. J Power Sources 195(13):4332–4337. doi:10.1016/j.jpowsour.2010.01.022

Zhang Y, Zhao Y, Bakenov Z, Gosselink D, Chen P (2014) Poly(vinylidene fluoride-co-hexafluoropropylene)/poly(methylmethacrylate)/nanoclay composite gel polymer electrolyte for lithium/sulfur batteries. J Solid State Electrochem 18(4):1111–1116. doi:10.1007/s10008-013-2366-y

Zheng JP, Liang RY, Hendrickson M, Plichta EJ (2008) Theoretical energy density of Li-air batteries. J Electrochem Soc 155(6):A432–A437. doi:10.1149/1.2901961

Zheng J, Lv D, Gu M, Wang C, Zhang J-G, Liu J, Xiao J (2013a) How to obtain reproducible results for lithium sulfur batteries? J Electrochem Soc 160(11):A2288–A2292. doi:10.1149/2.106311jes

Zheng J, Gu M, Chen H, Meduri P, Engelhard MH, Zhang J-G, Liu J, Xiao J (2013b) Ionic liquid-enhanced solid state electrolyte interface (SEI) for lithium-sulfur batteries. J Mater Chem A 1(29):8464–8470. doi:10.1039/C3TA11553D

Zheng G, Lee SW, Liang Z, Lee HW, Yan K, Yao H, Wang H, Li W, Chu S, Cui Y (2014) Interconnected hollow carbon nanospheres for stable lithium metal anodes. Nat Nanotechnol 9 (8):618–623. doi:10.1038/nnano.2014.152

Zhou G, Pei S, Li L, Wang D-W, Wang S, Huang K, Yin L-C, Li F, Cheng H-M (2014) A graphene–pure-sulfur sandwich structure for ultrafast, long-life lithium-sulfur batteries. Adv Mater 26(4):625–631. doi:10.1002/adma.201302877

Zu C, Manthiram A (2014) Stabilized lithium-metal surface in a polysulfide-rich environment of lithium-sulfur batteries. J Phys Chem Lett 5(15):2522–2527. doi:10.1021/jz501352e

Zu C, Su Y-S, Fu Y, Manthiram A (2013) Improved lithium-sulfur cells with a treated carbon paper interlayer. Phys Chem Chem Phys 15(7):2291–2297. doi:10.1039/C2CP43394J

Chapter 5
Perspectives

Li metal is an ideal anode material for rechargeable batteries, including Li-air, Li-S, and Li metal batteries using intercalation compounds or conversion compounds as cathode materials. Unfortunately, Li dendrite growth and low CE during the charge/discharge processes have largely prevented the use of Li metal as an anode for rechargeable batteries. To date, partial solutions for the prevention of Li dendrite growth have been identified, although they are only effective under certain conditions. For example, the PEO-based block copolymer electrolytes developed by Balsara et al. (2008, 2009) can prevent dendrite growth by using the strong shear strength of block copolymers, but these copolymers must work at elevated temperatures (~ 80 °C) when Li salt-PEO has acceptable conductivities. Selective electrolyte additives can prevent dendrite growth by a physical blocking mechanism or self-healing mechanism (Matsuda 1993; Aurbach et al. 2002b; Yoon et al. 2008; Aurbach and Chusid (Youngman) 1993), but most of these additives can prevent dendrite growth only at limited current densities. Too large a current density will induce a large voltage drop, which may force the additives to be deposited with Li. Most of these additives will be consumed during the cycling, so their effect will decrease with cycling iterations. On the other hand, improving the CE of Li cycling is a more challenging task. In this regard, the use of selected high concentration electrolytes seems to be a very promising approach that may lead to a CE of more than 99 % without dendrite formation (Qian et al. 2015b, 2016; Yuki Yamada and Yoshitaka Tateyama 2014). Although we are still a long way from realizing a superconcentrated electrolyte in practical application, we believe that this particular electrochemical feature and its detailed mechanisms will be of great value to the development of rechargeable Li metal batteries.

Generally speaking, high CE is a highly fundamental criterion for stable cycling of the Li metal anode. To have a high CE, side reactions between freshly/native deposited Li and electrolyte must be minimized. These reactions are proportional to the chemical and electrochemical activity of native Li when they are in direct

© Springer International Publishing Switzerland 2017
J.-G. Zhang et al., *Lithium Metal Anodes and Rechargeable
Lithium Metal Batteries*, Springer Series in Materials Science 249,
DOI 10.1007/978-3-319-44054-5_5

contact with surrounding electrolyte. They are also proportional to the surface area of deposited Li. Therefore, a high CE of Li cycling is usually a direct result of low reactivity between freshly deposited Li and the electrolyte, as well as of a low surface area of deposited Li. By contrast, a dendritic Li deposition always has a high surface area. This means that high CE Li deposition/stripping always correlates with low surface area Li deposition and suppressed Li dendrite growth. A stable CE value during long-term cycling also means that the SEI layer formed during Li deposition is relatively stable and that formation of new SEI layers is very minimal during each cycle. On the other hand, some electrolytes can foster dendrite-free Li deposition, but exhibit a CE of less than 80 % (Qian et al. 2015a; Ding et al. 2013). Therefore, the enhancement of CE is a fundamental factor in determining long-term, stable cycling of Li metal anodes.

To enable broad application of the Li anode, further fundamental studies need to be conducted to simultaneously address the two barriers discussed above. Future work also needs greater focus on addressing the cause of the problems instead of their consequences. The causes of Li dendrite formation and growth are the Li^+ concentration gradient during the charge/discharge process and nonuniform surface deposition. The origin of low CE is thermodynamic incompatibility between fresh Li metal and the electrolyte. The results of modeling and simulations suggest that an electrolyte with high ionic conductivity and a Li^+ transference number close to unity can dramatically flatten the concentration gradient, extending the "Sand's time," i.e., delaying or even stopping, Li dendrite formation. High shear modulus is also an important property for a functional electrolyte to suppress the Li dendrite growth.

In an effort to clearly understand the mechanism of Li dendrite formation and growth, various techniques have been adopted to characterize the morphology, components and structures of Li surfaces. Although some ex situ techniques provide much useful information, Li deposition/stripping is an interfacial reaction, so ex situ characterizations might destroy its original fine morphology. To understand the real dendrite formation and SEI layer formation, as well as their interactions, in situ or operando techniques (including operando TEM, SEM, AFM, or in situ optical techniques with high magnifications) can provide more information about local concentration gradient, Li dynamic nucleation and dendritic growth. They can also be used to observe the formation of dendrites and SEI layers at the same time. This is critical to finding an ultimate solution to overcoming the two barriers that have prevented the application of the Li anode to date.

Many modeling studies have been performed and have led to several critical conclusions that guided development of Li metal batteries. However, most of this modeling work has focused on the physical properties of dendrite growth. Most of the significant progress on increasing the CE of Li deposition/stripping [especially Aurbach's work (Aurbach et al. 1987, 1997, 2002a)] has been achieved by experimental efforts. To enable the application of Li metal anodes, the chemical reactions between Li metal and electrolytes (including solvents, salts and additives) need to be modeled and/or simulated to get a better understanding of the formation of the SEI layer and its chemical stability, ionic conductivity, and mechanical flexibility. Furthermore, the interactions between Li dendrites and the SEI layer need to be better understood and incorporated into the simulation process to get a realistic picture of the Li deposition/stripping process. For example, an SEI layer with limited flexibility will be able to tolerate small variations on the Li surface. This will prevent repeated breakdown and reformation of the SEI layer, which is one of the main sources of low CE during a Li anode's operation.

Li is thermodynamically unstable with any organic solvent; they react instantaneously to form an SEI layer. Therefore, the stability and flexibility of the SEI layer is the most critical factor determining the performance of a Li metal anode, especially its CE. The components of the electrolytes, including solvents, Li salts, and additives determine the chemical composition, ionic conductivity, and mechanical properties of the SEI layer. Compared to liquid electrolytes, polymer materials (especially PEO) are much more compatible with fresh Li. Therefore, PEO and its derivatives have been investigated extensively for their applications to rechargeable Li metal batteries over the last three decades. However, the PEO electrolyte still cannot prevent Li dendrite growth because it has a relatively low ionic conductivity and a small Li^+ transference number, causing a steep Li^+ concentration gradient during polarization, as anticipated by modeling. An effective strategy is to develop single-ion conductors with a unity Li^+ transference number, plus a strong shear modulus with block copolymers, which greatly reduce the concentration gradient and effectively retard Li dendrite formation and growth. Another possible solution is to combine the self-healing mechanism with an electrolyte that can form a stable SEI layer with Li. Further improvement in the conductivity of the electrolyte can reduce the internal voltage drop under high current density conditions. Development of inorganic, solid-state Li^+ ion conductors with good mechanical strength and stability and high Li ionic conductivity as well as excellent compatibility with Li metal could present an ideal solution for rechargeable Li metal batteries.

Creating practical applications of Li metal anodes for rechargeable batteries may require a combination of different approaches. A well balanced Li deposition/stripping rate is also critical for long term stability of Li cycling. It is likely that no single ideal solution will make a Li anode work well under all conditions, so different solutions may have to be designed for different applications. Although there are still many obstacles to overcome, we are optimistic that Li metal can be used as an anode in rechargeable batteries in the near future. This will enable wide applications of rechargeable Li-S batteries, Li-air batteries, and Li metal batteries using intercalation compounds as cathode.

References

Aurbach D, Chusid (Youngman) O (1993) In situ FTIR spectroelectrochemical studies of surface films formed on Li and nonactive electrodes at low potentials in Li salt solutions containing CO_2. J Electrochem Soc 140(11):L155–L157

Aurbach D, Daroux ML, Faguy PW, Yeager E (1987) Identification of surface films formed on lithium in propylene carbonate solutions. J Electrochem Soc 134(7):1611–1620

Aurbach D, Zaban A, Ein-Eli Y, Weissman I, Chusid O, Markovsky B, Levi M, Levi E, Schechter A, Granot E (1997) Recent studies on the correlation between surface chemistry, morphology, three-dimensional structures and performance of Li and Li-C intercalation anodes in several important electrolyte systems. J Power Sources 68:91–98

Aurbach D, Zinigrad E, Cohen Y, Teller H (2002a) A short review of failure mechanisms of lithium metal and lithiated graphite anodes in liquid electrolyte solutions. Solid State Ionics 148:405–416

Aurbach D, Zinigrad E, Teller H, Cohen Y, Salitra G, Yamin H, Dan P, Elster E (2002b) Attempts to improve the behavior of Li electrodes in rechargeable lithium batteries. J Electrochem Soc 149(10):A1267–A1277. doi:10.1149/1.1502684

Balsara N (2008). http://www.1.eere.energy.gov/vehiclesandfuels/pdfs/merit_review_2008/exploratory_battery/merit08_balsarapdf

Balsara NP, Singh M, Eitouni HB, Gomez ED (2009). US Patent Appl No 0263725 A1

Ding F, Xu W, Graff GL, Zhang J, Sushko ML, Chen X, Shao Y, Engelhard MH, Nie Z, Xiao J, Liu X, Sushko PV, Liu J, Zhang J-G (2013) Dendrite-free lithium deposition via self-healing electrostatic shield mechanism. J Am Chem Soc 135(11):4450–4456. doi:10.1021/ja312241y

Matsuda Y (1993) Behavior of lithium/electrolyte interface in organic solutions. J Power Sources 43–44:1–7

Qian J, Xu W, Bhattacharya P, Engelhard M, Henderson WA, Zhang Y, Zhang J-G (2015a) Dendrite-free Li deposition using trace-amounts of water as an electrolyte additive. Nano Energy 15:135–144. doi:10.1016/j.nanoen.2015.04.009

Qian J, Henderson WA, Xu W, Bhattacharya P, Engelhard M, Borodin O, Zhang JG (2015b) High rate and stable cycling of lithium metal anode. Nat Commun 6:6362. doi:10.1038/ncomms7362

Qian JF, Adams BD, Zheng JM, Xu W, Henderson WA, Wang J, Bowden ME, Xu SC, Hu JZ, Zhang JZ (2016) Anode-free rechargeable lithium metal batteries. Adv Funct Mater 2016. doi:10.1002/adfm.201602353

Yoon S, Lee J, Kim S-O, Sohn H-J (2008) Enhanced cyclability and surface characteristics of lithium batteries by Li-Mg co-deposition and addition of HF acid in electrolyte. Electrochim Acta 53(5):2501–2506. doi:10.1016/j.electacta.2007.10.019

Yamada Y, Furukawa K, Sodeyama K, Kikuchi K, Yaegashi M, Tateyama Y, Yamada A (2014) J Am Chem Soc 136:5039–5046

Index

© Springer International Publishing Switzerland 2017
J.-G. Zhang et al., *Lithium Metal Anodes and Rechargeable*
Lithium Metal Batteries, Springer Series in Materials Science 249,
DOI 10.1007/978-3-319-44054-5

Printed in the United States
By Bookmasters